I0120330

ETERNAL ENERGY

A New Blueprint for Sustainable Power and Global Equity

US Provisional Patent Application # 63/715,475
Filed November 1, 2024

MICHAEL ELFELLAH

AIMQWEST BOOKS
AIMQWESTBOOKS.COM

First Edition

Publication Date: November 1, 2024

ISBN: 979-8-992-17620-9

Foreword

In a world driven by an unyielding demand for energy, humanity stands at a crossroads. **Eternal Energy**: A New Blueprint for Sustainable Power and Global Equity is not merely a book about advancing power systems or mitigating shortages; it is a bold proposition for a future where energy is not a resource to be hoarded, rationed, or fought over, but a force that amplifies itself—a boundless entity that powers growth and nurtures global equity.

This work embarks on a quest beyond the familiar paradigms of fossil fuels, renewable intermittency, and the geopolitical tensions they create. It envisions energy as an ever-renewing resource that transcends scarcity and defies the constraints of conventional power models. By blending cutting-edge science with principles of equity and sustainability, this book proposes a future of "compounding energy," where power perpetuates itself, cycling back in regenerative, self-sustaining systems that can fuel human civilization indefinitely.

Central to this journey is the concept of energy compounding, a revolutionary model in which each unit of energy contributes to the creation of more. This vision transcends terrestrial limitations through advancements in zero-gravity physics, superconductivity, and space-based power systems. These developments reframe energy not merely as a consumable commodity but as a dynamic force that, much like a natural ecosystem or the financial principle of compounding interest, thrives and grows.

Yet, this work does not stop at technicalities; it places ethics, equity, and global welfare at its core. It challenges the world's leaders, scientists, and citizens to reevaluate how we allocate, consume, and govern power. It imagines an infrastructure that extends beyond profit-driven monopolies, offering a model where energy fosters equality, lifts societies, and enables individuals to thrive.

If the ambitions of Eternal Energy seem vast, it is because this book is an invitation to reimagine humanity's relationship with power itself. It calls for a vision of progress that is as compassionate as it is advanced. This work offers not only innovation but a roadmap for a sustainable, equitable future. It is a book that dares to suggest that humanity's energy future can be one of limitless, ethical abundance.

To all readers, policymakers, and pioneers: may you find within these pages the inspiration to build a world where energy empowers, uplifts, and unites us all.

Michael K Elfellah, Founder of AIMQWEST Corporation and author of **Eternal Energy**: A New Blueprint for Sustainable Power and Global Equity, Plano, Texas — November 2024
"For a world where energy serves every soul."

TABLE OF CONTENTS

Preface: The Call to a New Energy Era

Since the dawn of civilization, energy has underpinned humanity's journey, its triumphs, and its profound vulnerabilities. The flickering fires of early man transformed nights into havens of warmth and safety, marking the first human strides in harnessing nature's forces. Fire, as simple and finite as it was, signified more than survival—it symbolized humanity's endless pursuit to bend nature to its needs. From that primitive mastery, humanity embarked on an evolving quest to control, amplify, and reshape energy, with each era marking a transformative milestone.

As the cradle of ancient civilizations, agriculture thrived on the muscle of humans and animals, a humble but powerful use of organic energy. This exertion of direct force irrigated fields, constructed the wonders of antiquity, and enabled the foundational growth of societies. Yet, every advance bore the limitations of its time. The Greeks and Romans pioneered the early use of wind and water mills, the grinding sounds of which echoed a pivotal shift in human industry. These were the first murmurs of power derived from natural forces, whispering promises of efficiency that ancient muscle could not provide. Still, the methods were bound by nature's constraints, tethered to resources that were as finite as they were essential.

With the advent of the Industrial Revolution, humanity experienced an energy renaissance—one where coal lit the fires of factories and propelled the gears of machinery, transforming landscapes, societies, and economies. The steam engine stood at the heart of this era, a machine that epitomized human ingenuity and catalyzed monumental shifts in production and transportation. From wood to coal, societies pursued energy sources that offered increased density, predictability, and power. Yet, the increased reliance on coal marked the beginning of an environmental debt that modern civilization is still striving to repay. The rise of fossil fuels brought immense growth but came with a price—the first shadow of scarcity amid abundance, as every ton of coal burned foretold eventual depletion and harm.

As humanity crossed into the 20th century, oil and natural gas rose to dominance, enabling unprecedented levels of mobility and convenience. Oil became the lifeblood of modernity, fueling vehicles, planes, and industrial advancements, each step propelling society closer to a global economy interconnected by an unyielding thirst for fuel. The electric grid, powered initially by coal and then diversifying to hydropower, nuclear, and gas, transformed human life, extending energy's reach to illuminate cities and connect communities. Yet, the finite nature of fossil fuels, coupled with their devastating environmental impacts, began to pose a grim question: could civilization sustain itself on borrowed time, depending on resources that strained the earth's ecosystems and fragile climate?

In response, a wave of renewable initiatives emerged, marking the late 20th and early 21st centuries with the promise of sustainability. Solar panels and wind turbines began dotting landscapes, heralding a cleaner, greener era. Governments, industries, and citizens alike invested in renewable technology, driven by the imperative to reduce reliance on fossil fuels. Solar, wind, geothermal, and biofuels became rallying points, offering pathways to a sustainable future. However, the technological and infrastructural limitations of renewables soon revealed themselves. Intermittency in energy production, challenges in storage, and high initial costs underscored the complexity of moving toward a fully renewable world. The cycles of the sun and wind are bound to nature, often failing to match the demands of society's energy needs at any given moment. Advances in battery technology and grid management offer progress, but humanity continues to struggle with the balancing act of demand, sustainability, and efficiency.

Thus, the journey of energy, from primitive fire to modern renewables, stands as both a testament to human ingenuity and a sobering reminder of natural constraints. Every leap—from coal to electricity, from oil to solar—has advanced civilization while illuminating the limitations imposed by resources and technology. In facing the realities of finite resources and environmental degradation, humanity has arrived at a crossroads. The concept of energy scarcity—a shadow cast over every advance—demands an answer that transcends current paradigms, one that reimagines energy not merely as a resource to consume, but as a force to renew, grow, and sustain. It is a vision no longer relegated to the bounds of imagination, as breakthroughs in science and technology now edge us closer to a world where energy scarcity may be a relic of the past.

Yet, standing on the brink of a transformative era, modern society must confront its paradox. While technology has woven energy into the fabric of everyday life, enabling everything from microprocessors in personal devices to grids that power entire cities, the very systems that sustain us face unprecedented constraints. Fossil fuels, long the bedrock of industrial growth, pose environmental threats, while renewables, though promising, struggle to fully meet global demands. The dynamic tension between energy abundance and scarcity creates a backdrop of urgency—a precipice that calls for innovative solutions beyond traditional frameworks.

As humanity seeks a new chapter in its energy narrative, it faces the dual imperatives of achieving abundant power while protecting the planet. Fossil fuel extraction is fraught with environmental costs: deep-sea drilling, fracking, and mining irreparably harm ecosystems, polluting vital water sources and endangering biodiversity. The byproduct of this dependence—carbon emissions—accelerates climate change, raising questions about the sustainability of current practices. Renewable energy, meanwhile, though heralded as a solution, poses its own challenges. Solar and wind energy, while abundant, are inherently tied to nature's rhythms, constrained by day, night, and the patterns of wind. To meet energy needs consistently, society must invest heavily in storage solutions, though current battery technology falls short of ensuring round-the-clock supply. The high demands for lithium, cobalt, and other materials needed for these storage solutions add a new layer of environmental strain, highlighting that even green energy comes with its costs.

This drive for a stable, clean, and universally accessible energy source transcends mere technical aspiration; it is an existential necessity, one that presses humanity to confront the limitations of fossil fuels, the promises and pitfalls of renewables, and the need for a new approach that reshapes our relationship with power itself. To create a resilient, abundant, and sustainable energy system, humanity must think beyond renewable substitutions and toward revolutionary solutions that redefine energy's role within society. This vision is more than a dream; it is an imperative that calls for exploration, innovation, and a new paradigm that aligns technological capabilities with sustainable growth, liberating civilization from the shackles of scarcity.

Despite advancements, modern energy infrastructure faces a conundrum. As global demand grows insatiably, powered by everything from vast industrial machinery to handheld devices, the production, storage, and distribution of energy teeter on precarious boundaries. Fossil fuels—long celebrated for their energy density and reliability—have become synonymous with pollution, finite reserves, and complex geopolitics. Meanwhile, renewables, despite their promise, struggle with the inherent unpredictability and limitations of nature. Solar and wind power, for example, hinge on conditions beyond human control, with production fluctuating with every passing cloud or shift in the wind. Consequently, renewable energy requires a delicate balance of investment in infrastructure and innovation to bridge the gap between idealistic aspirations and practical applications.

As the world commits to reducing carbon emissions, the allure of renewables has shifted from an alternative to an imperative. Governments and private sectors alike champion solar, wind, geothermal, and hydropower, with the intent to not only curb reliance on fossil fuels but also restore a semblance of environmental balance. Yet, the logistical and technological challenges inherent in renewables present a sobering truth.

Intermittent and variable, renewable sources cannot, in their current forms, wholly sustain the relentless demands of modern life. Efforts to address this through battery storage—whether via lithium-ion or emerging technologies like solid-state and hydrogen fuel cells—offer some respite. However, these solutions bring their own environmental and economic complexities. Mining for lithium, cobalt, and other rare elements necessary for large-scale batteries places enormous strain on ecosystems, disrupts local communities, and introduces a new dependency on regions with these critical resources. The cycle of extraction and consumption continues, albeit with a greener guise, underscoring the limitations of existing renewable infrastructure to provide a truly sustainable and universally accessible energy supply.

This reality presses upon the world the urgency of a more radical reimagining of energy. Current solutions, while critical steps in reducing fossil fuel dependency, remain bound by physical constraints that tether society to a scarcity-driven paradigm. The vision of an abundant energy future proposes an entirely different approach, one that liberates humanity from these limitations by redefining the foundational principles of power generation. Imagine a world where energy not only meets immediate needs but expands its capacity with every cycle—a self-sustaining, self-compounding system. This vision, once relegated to science fiction, is edging closer to possibility through groundbreaking advances in zero-gravity environments, superconductivity, and space-based power generation.

The essence of this vision lies in energy compounding, an approach inspired by natural cycles and financial principles. Just as ecosystems thrive by recycling resources in an unbroken loop, or as compounded interest in finance grows exponentially, energy compounding aims to create a system where each unit of power sustains itself while generating a surplus. This surplus would be reinvested back into the system, creating a dynamic feedback loop that amplifies energy availability over time. By shifting from a model of consumption to one of regeneration, energy compounding challenges the scarcity-bound thinking that has long governed human progress. Instead, it offers a path toward sustainable abundance, where power is no longer a finite commodity but a resource that grows in tandem with society's needs.

To realize energy compounding on a practical scale, several scientific and environmental breakthroughs are essential. The setting for such an ambitious energy system transcends terrestrial limitations, calling for conditions only achievable in the vast, frictionless expanse of space. In a zero-gravity environment, machinery can operate without the drag and gravitational pull that, on Earth, constantly consume energy. By positioning energy stations in stable orbit, humanity could harness the advantages of zero gravity to maintain a compounding cycle of power generation with minimal energy loss. At Lagrange points—regions in space where gravitational forces between Earth and other celestial bodies reach equilibrium—a motor-generator system could operate with near-perfect efficiency. Free from terrestrial resistance, it would maximize every cycle's output, generating surplus energy that feeds back into the system.

However, zero-gravity alone does not achieve the compounding effect; it must be paired with the revolutionary properties of superconductivity. First observed in the early 20th century, superconductivity is a state wherein materials conduct electricity without resistance when cooled to ultra-low temperatures. Resistance, which converts part of the energy in conventional systems into heat, thereby diminishing efficiency, is eliminated in superconductors. In the context of an energy-compounding system, superconductors allow for the uninterrupted flow of current, significantly reducing losses and enabling power to circulate through the system unimpeded. Imagine a motor-generator that, instead of consuming additional power to counteract resistance, continually reinvests its output, compounding with each cycle to create an ever-growing reserve of energy.

By leveraging both zero-gravity and superconductivity, the compounding model presents an unprecedented level of efficiency. In this design, energy becomes an expanding resource, scaling alongside human ambitions without the conventional losses and limitations of traditional systems. What emerges is a system that could redefine global infrastructure, one capable of generating, storing, and distributing power on a scale that meets, and even exceeds, humanity's needs.

For this vision to materialize on Earth, the challenge of energy transmission must also be addressed. The idea of generating energy in space and transmitting it back to Earth has captivated scientists for over a century. Nikola Tesla, whose groundbreaking work in wireless energy transmission inspired generations of scientists, proposed a world where power could be sent across great distances without physical connections. Today, advancements in resonant inductive coupling—a method that allows energy to jump from one coil to another without wires—have brought Tesla's vision closer to realization. A space-based energy-compounding station, utilizing this technology, could theoretically beam power directly to Earth, where it would be captured by receivers stationed across the globe. In this model, the energy grid as we know it would be transformed, evolving from a terrestrial network of interconnected infrastructure to a dynamic system that draws power from the cosmos itself.

Such a system promises to address some of humanity's most pressing challenges, from energy inequality and economic disparity to environmental degradation. With an inexhaustible supply of clean energy, society could finally transcend the cycles of scarcity that have historically dictated resource distribution. Nations that have long struggled with limited energy access would gain the power to grow and develop, unimpeded by the costs and constraints of traditional energy systems. Industries could operate sustainably, empowered by an energy infrastructure that does not strain resources or contribute to pollution. With the right frameworks in place, energy compounding could become the cornerstone of a more equitable and sustainable global economy, creating opportunities for innovation, growth, and progress that reach every corner of the world.

The implications of energy abundance go far beyond technology and economics. With the ability to meet fundamental needs without exhausting finite resources, humanity could shift its focus from survival to stewardship. Water scarcity, for example, would be alleviated as desalination plants, powered by limitless energy, transform seawater into fresh water for even the driest regions. Food production would flourish through energy-intensive vertical farming, enabling year-round cultivation and reducing dependence on arable land. This abundant energy supply would allow entire societies to develop sustainably, without the environmental compromises often tied to economic growth.

Perhaps most profoundly, the vision of energy abundance allows humanity to become a true steward of the planet. No longer bound by the need to exploit natural resources, societies could turn their energies toward restoring ecosystems, preserving biodiversity, and even reversing the effects of climate change. Projects such as large-scale reforestation, habitat restoration, and carbon capture would gain unprecedented support, creating a harmonious balance between technological progress and environmental preservation. In this world, energy abundance is not merely a source of power but a catalyst for healing and growth, enabling humans to coexist with nature in a way that sustains both.

In this transformative vision, the promise of energy compounding is not just about efficiency; it is a redefinition of how humanity perceives energy itself. By shifting from a paradigm of scarcity to one of sustainable abundance, civilization would no longer view energy as a constraint but as a boundless force for good. In such a world, energy would fuel not only physical infrastructure but also the aspirations of generations to come, empowering individuals, communities, and nations to reach their fullest potential.

As humanity steps into this new era, it is invited to embrace the possibilities of energy compounding, to reimagine its relationship with power and progress, and to envision a future where energy scarcity is a relic of history. In doing so, society stands on the cusp of a revolution that transcends technology, touching the very essence of what it means to live in harmony with the planet and with one another.

The path to energy abundance did not materialize in isolation; it draws from centuries of insight, from the regenerative systems of nature to the principles of financial growth. Nature, a master of cyclical energy flows, provides an elegant blueprint of sustainability where resources are continuously recycled, and nothing is ever truly lost. In the natural world, energy circulates through ecosystems in an unbroken loop, passing from plants to herbivores to carnivores and, eventually, back into the earth, where it nourishes life anew. This balanced exchange exemplifies a form of energy compounding that has long sustained Earth's biodiversity. In a similar way, humanity has observed and sought to replicate nature's seamless regeneration, with early thinkers wondering if energy systems might likewise generate and reinvest surplus.

Nature's cycles parallel another powerful concept: the compounding of interest in finance. The principle that a modest investment can grow exponentially if returns are reinvested has, for centuries, illustrated the transformative potential of disciplined accumulation. Financial compounding, celebrated as a source of wealth creation, demonstrates how a process of continuous reinvestment can turn modest beginnings into substantial reserves over time. Albert Einstein famously described compound interest as the "eighth wonder of the world," underscoring its potential to turn incremental gains into exponential growth. In the realm of energy, this principle suggests a similar possibility—a system in which each surplus of energy is reinvested, creating a cascading effect of power production.

This blend of natural wisdom and financial insight forms the foundation of energy compounding, an approach that envisions energy systems as capable of self-sustaining growth. Rather than focusing on merely meeting present demand, this model proposes an infrastructure designed to expand over time, continuously increasing its capacity by reinvesting each unit of surplus energy. This regenerative energy model mirrors the closed-loop systems found in ecosystems, where energy flows without depletion. By designing such a system, humanity could finally break free from the cycles of scarcity, embracing a future where energy no longer limits progress but empowers it.

Throughout history, the notion of energy surplus and reinvestment has inspired thinkers and inventors alike, sparking a vision of a self-sustaining system that could not only meet human needs but amplify its capacity indefinitely. Engineers in the 19th century, amidst the industrial fervor, observed how machines and engines often fell short of their theoretical potentials, limited by inefficiencies, friction, and energy losses. Yet, these limitations ignited a quest for mechanisms that could capture and minimize these losses, driving the development of more efficient engines and turbines. The goal was to design a system where energy could, theoretically, exceed input requirements, generating a surplus that could sustain and grow.

At the turn of the 20th century, advances in thermodynamics, a field focused on understanding the relationships between heat, energy, and work, offered new insights into energy management. Engineers and scientists began exploring how systems could be optimized to not only perform efficiently but also generate surplus power. Although constrained by technological limitations of the time, these early explorations laid a foundation for modern theories of energy compounding. As society entered the era of financial growth, the idea of "returns on returns" offered an intriguing framework for how energy, too, might follow a compounding model, expanding exponentially if its output could reinvest into itself.

As the 20th century progressed, scientists turned their attention to new environments that could accommodate a truly efficient, compounding energy system. The discovery of superconductivity, which demonstrated that certain materials could conduct electricity without resistance when cooled to ultra-low temperatures, unlocked new possibilities. Superconductors operate with near-perfect efficiency, allowing electricity to flow indefinitely without loss. This breakthrough offered a tantalizing glimpse into the efficiencies required for energy compounding. With materials that could conduct power without dissipating energy as heat, scientists began envisioning an energy infrastructure that could sustain itself with minimal input, mirroring the compounding process seen in ecosystems and finance.

Superconductivity's potential led researchers to explore the extreme conditions under which these materials thrived, particularly the frigid, zero-gravity environments of space. Zero gravity, by minimizing friction and eliminating the forces that typically degrade machinery, offers an ideal setting for maximizing efficiency. In a gravitational vacuum, a motor-generator could theoretically maintain continuous operation with reduced energy input, preserving momentum and reinvesting surplus energy. Here, the concept of energy compounding takes on practical significance, as the absence of gravity and friction allows systems to operate without the drag and resistance that constrain Earth-bound technologies.

In pairing zero-gravity conditions with superconductivity, scientists began envisioning an energy infrastructure capable of compounding power in a sustainable, continuous cycle. This concept, once confined to theoretical research, now appears increasingly viable in a world where space-based technology and advanced materials science offer new frontiers. With superconductors operating in the frictionless, low-temperature conditions of space, an energy system could achieve efficiencies unimaginable within Earth's atmosphere, creating an energy surplus capable of reinvestment. Such a system promises not only to meet humanity's energy needs but to scale beyond them, delivering abundant power to fuel civilization's highest aspirations.

As research advances, the synthesis of these scientific breakthroughs brings the vision of energy compounding within reach. By marrying the efficiencies of superconductors with the stability of zero-gravity environments, scientists can design an energy system that mirrors the regenerative cycles of nature and the exponential growth of finance. These discoveries invite society to reimagine energy as a self-sustaining force—a dynamic system capable of powering itself and reinvesting its output to create a cycle of abundance. This vision shifts the very notion of power generation from one of extraction and consumption to one of reinvestment and expansion, laying the groundwork for a future where energy scarcity no longer dictates the bounds of human potential.

The implications of energy abundance are transformative, and realizing this vision calls for technology capable of transcending Earth's physical limits. By establishing energy-compounding stations in space, humanity can harness the efficiencies of zero gravity and superconductivity to create a system with near-perfect energy retention. In these orbital outposts, energy could be generated, stored, and even beamed back to Earth through advanced transmission methods, ensuring that communities worldwide have access to consistent, reliable power. This approach marks a fundamental shift, turning space from a distant frontier into an active resource that feeds directly into Earth's energy ecosystem.

Wireless energy transmission, inspired by the pioneering theories of Nikola Tesla, plays a crucial role in bridging the gap between space-generated power and terrestrial grids. Tesla's groundbreaking work in resonant inductive coupling suggested that energy could travel without physical connections, a concept that has become increasingly relevant in the context of space-based energy systems. Advances in this field now allow for the potential to beam energy across vast distances with minimal loss, enabling a direct link between space-based generators and receivers on Earth. Such a transmission system would redefine the concept of an energy grid, expanding its reach from a network of wires and substations to a truly global infrastructure.

The strategic placement of these energy-compounding stations in stable orbits, such as Lagrange points, enables continuous operation without the need for frequent adjustments. At these points, the gravitational forces of Earth and other celestial bodies are balanced, creating zones of stability ideal for long-term installations. By positioning energy stations at these points, humanity can build a resilient and scalable infrastructure that maximizes the benefits of zero-gravity and superconductive environments. This self-sustaining network of energy compounding would not only alleviate the immediate constraints of terrestrial systems but also offer a model for global distribution that transcends the limitations of geography.

The impact of an abundant energy system stretches far beyond technical advancements; it reshapes society itself. In a world powered by self-sustaining energy, communities previously marginalized by limited access to resources could emerge as thriving centers of growth. Energy poverty, a barrier to economic development and quality of life, would be significantly reduced, as abundant power enables advancements in healthcare, education, and infrastructure. Entire industries could reimagine their operations without the environmental and economic costs tied to traditional energy consumption, creating a ripple effect of sustainable progress across sectors.

The vision of energy compounding not only addresses immediate needs but empowers humanity to face its greatest challenges. In an energy-abundant world, issues such as water scarcity, food security, and climate change become more manageable. For instance, desalination plants powered by this limitless energy could transform seawater into drinkable resources, alleviating water shortages in arid regions. Vertical farming, made viable by abundant power, could reduce dependency on arable land, ensuring a reliable food supply that sustains populations without expanding into sensitive ecosystems. This model of energy abundance aligns human advancement with environmental protection, fostering a future where progress and preservation coexist harmoniously.

Ultimately, the vision of energy compounding invites humanity to think beyond survival and toward stewardship. With energy no longer a limiting factor, society could invest in large-scale environmental restoration projects, from reforestation to carbon capture, allowing for active intervention in climate stability. The energy to reverse environmental degradation, once thought of as an impossible resource, becomes readily available, enabling humanity to restore balance to ecosystems and protect biodiversity. This abundance transforms energy from a source of consumption to a means of healing, reorienting technological progress toward sustaining and nurturing the natural world.

The energy-compounding model does more than eliminate scarcity; it redefines how humanity relates to power and resources. This vision presents energy as a force that grows in tandem with society's needs, fostering a sustainable relationship between human achievement and planetary health. By embracing a regenerative energy model, civilization opens the door to a future where technology and nature coexist, where the potential for human growth is matched by a commitment to environmental resilience. In this world, energy abundance serves not as a mere enabler of progress but as a bridge to harmony, shaping a legacy of sustainable innovation that reaches across generations.

The vision of energy compounding, when fully realized, offers a blueprint for sustainable progress that touches every facet of society. No longer would energy be a resource rationed and reserved; it would become a catalyst for growth, universally accessible and continuously renewable. At its core, this blueprint represents a paradigm shift, urging humanity to transcend scarcity-driven models and adopt a regenerative approach that fuels both human ambition and planetary stewardship.

At the center of this blueprint is the commitment to equity. Energy scarcity has historically created an imbalance, with access to power defining the progress and prosperity of nations, communities, and individuals. In many parts of the world, limited access to affordable energy inhibits economic growth, education, and healthcare, keeping entire populations in cycles of poverty. This disparity has been deeply rooted in the existing energy infrastructure, where distribution networks favor certain regions while bypassing others. The vision of an energy-compounding infrastructure challenges this inequality, proposing a model where energy abundance is shared, transforming the lives of those who have long been marginalized by limited resources. An energy-compounding system, by providing a steady and growing supply of power, redefines progress as inclusive, accessible, and universal.

To bring this vision to life, it is essential to establish energy-compounding stations that operate at near-perfect efficiency in space. Located at stable Lagrange points, these stations leverage zero-gravity conditions and superconductive technology to produce energy continuously, free from the losses and inefficiencies that plague terrestrial systems. Once generated, this energy can be transmitted wirelessly to Earth, where it would be received by strategically placed infrastructure, allowing even the most remote regions to access power. This global reach shifts energy distribution from a privilege to a right, enabling equitable growth and development.

This blueprint also embodies a new economic model where the extraction and depletion of resources are no longer the drivers of progress. In place of finite energy reserves that require constant replenishment, humanity gains a system that compounds in capacity, growing with each cycle and eliminating the need for resource-based competition. Energy becomes a resource that empowers, rather than limits, human achievement. Industries would no longer need to expend resources simply to meet energy demands; they could direct those resources toward innovation, sustainability, and social impact. The resulting model encourages economic structures based on growth, creativity, and environmental harmony, ushering in a future where scarcity-based competition gives way to cooperation and shared prosperity.

The societal impact of energy abundance extends beyond economics to influence culture, governance, and the environment. With the assurance of stable, renewable energy, humanity could fundamentally reframe its priorities, shifting from survival to collective flourishing. A world no longer preoccupied with energy scarcity would have the freedom to focus on education, scientific exploration, and environmental conservation. This paradigm invites society to adopt a broader perspective on progress, recognizing that true advancement involves not only technological innovation but also the ethical and equitable distribution of resources. As energy becomes a resource available to all, it empowers a more enlightened approach to governance, one that prioritizes transparency, accountability, and public welfare.

To guide the implementation of such a transformative system, a strong ethical foundation is required—one that safeguards against monopolization and ensures that energy abundance serves the common good. History is rife with examples of resources that, though abundant, were controlled by a few, leading to monopolies and exacerbated inequalities. In an era of energy abundance, it is imperative that no single entity, nation, or corporation has exclusive control over this resource. Governance structures must be designed to protect public access, prevent exploitation, and ensure that the benefits of energy compounding reach all communities.

This model of governance would involve international cooperation, where policies are established to manage, distribute, and protect energy resources as a shared asset. Global organizations, governments, and private sectors must come together to create regulatory frameworks that enforce fair practices, promote ethical innovation, and establish standards for responsible energy management. Such governance would be proactive, focusing on equitable distribution and access while maintaining transparency in decision-making processes. By aligning energy management with principles of social equity and environmental sustainability, humanity could redefine energy not merely as a commodity but as a universal right.

The blueprint for energy abundance calls for an inclusive model of progress, one that unites nations, communities, and individuals in the pursuit of a sustainable, harmonious future. Energy compounding, when managed with care and responsibility, has the potential to bridge historical divides, fostering a global society rooted in shared values and mutual respect. With energy abundance, humanity could transcend its current limitations, addressing not only economic disparities but also environmental degradation, fostering a balanced relationship between technological advancement and ecological health.

The environmental implications of this model are profound. With access to limitless, renewable power, society could accelerate environmental restoration efforts on a scale previously unimaginable. Large-scale projects like reforestation, ocean cleanup, and habitat restoration would become feasible, supported by an energy infrastructure that encourages, rather than restricts, these initiatives. In a world no longer reliant on fossil fuels, the air would clear, the waters would purify, and ecosystems could begin to heal. Energy abundance makes possible a level of environmental stewardship that redefines humanity's role on Earth, allowing society to transition from a force of extraction to one of restoration.

Moreover, the energy-compounding model would enable the development of advanced technologies for climate intervention and environmental protection. Innovations like carbon capture, bioengineering, and climate adaptation technologies could be deployed on a global scale, creating proactive solutions to combat climate change and preserve biodiversity. With the power to sustain such projects indefinitely, energy abundance transforms humanity from a passive observer of environmental decline to an active agent of renewal, supporting a vision of sustainable coexistence with the natural world.

As humanity embraces this energy-compounding blueprint, it is invited to consider the ethical responsibilities that accompany such power. Energy abundance is not merely a technological achievement; it is an opportunity to reshape societal values, prioritizing not only efficiency and growth but also justice, fairness, and collective well-being. This paradigm demands a reimagining of what it means to progress, challenging humanity to adopt a holistic perspective that honors the dignity of individuals, the health of ecosystems, and the long-term sustainability of civilization.

The shift toward energy abundance offers a chance to redefine prosperity in a way that aligns with ethical principles and environmental imperatives. Prosperity in this vision is not marked by unchecked consumption but by a balanced approach to resources, where human achievement is measured by its harmony with the planet. In a world where energy scarcity is no longer a concern, society could focus on cultivating innovation that supports resilience, sustainability, and inclusivity, fostering an economy that prioritizes both human and environmental welfare.

Ultimately, the energy-compounding vision is an invitation to dream of a future unburdened by scarcity, where humanity's greatest challenges—poverty, inequality, environmental degradation—become solvable within an infrastructure that supports both abundance and equity. This blueprint does not simply present a technical solution; it offers a transformative framework for redefining civilization's relationship with energy, the planet, and itself. By embracing a model of energy abundance, humanity could craft a legacy of progress that honors the principles of stewardship, sustainability, and shared prosperity, creating a world where the potential for growth is as limitless as the energy that fuels it.

As humanity contemplates this transformative shift, the responsibility of knowledge becomes paramount. In a world on the cusp of energy abundance, the insights gained from centuries of exploration, innovation, and environmental learning converge into a single, powerful realization: with great power comes great responsibility. The ability to harness limitless energy brings with it ethical obligations that must guide every aspect of its application, from technological development to social governance. The power to redefine civilization's trajectory must be wielded with a commitment to fairness, transparency, and sustainability, ensuring that the benefits of energy abundance serve not just a privileged few but the entire global community.

Knowledge in this new era must be shared, transparent, and accessible, promoting an open exchange of ideas and innovation that safeguards against misuse and monopolization. In this respect, the responsibility of knowledge mirrors the responsibility of stewardship. Just as energy must be managed for the collective good, so too must the insights, breakthroughs, and technologies associated with energy compounding be protected from exploitation. Scientific and technological advancements should be democratized, making the benefits of energy abundance available to all, from the wealthiest nations to the most underserved communities.

The commitment to knowledge-sharing is not only a moral imperative but a strategic one. In a world where energy abundance reshapes global dynamics, collaboration and transparency ensure that no single entity can dominate or control this resource. By fostering a spirit of open innovation and shared responsibility, humanity can mitigate the risks of monopolization and cultivate an infrastructure that benefits everyone. In this way, the vision of energy compounding becomes more than a technical achievement; it becomes a societal shift, a reimagining of how humanity defines and distributes power, not just in the physical sense but in the ethical and philosophical sense as well.

At the heart of this transformative vision is an invitation to engage, collaborate, and dream. The path to energy abundance is not only about harnessing technology but about reshaping society's values, challenging long-held assumptions, and fostering a mindset of shared stewardship. This vision invites humanity to imagine a world where energy fuels not only technological advancement but also human dignity, environmental preservation, and collective growth. It is an opportunity to transcend the limitations of the past and create a legacy that reflects humanity's highest ideals—justice, inclusivity, and harmony with the planet.

As society moves forward, each step in the journey toward energy abundance reinforces the need for a collective approach. This journey is not just about energy; it is about the future of civilization itself. In reimagining the role of energy, humanity redefines its potential, creating a world where power serves as a bridge to equity, a tool for environmental stewardship, and a foundation for lasting peace and prosperity. With this vision, the blueprint for energy abundance becomes a call to action, urging every individual, community, and nation to contribute to a world defined not by scarcity, but by shared hope, opportunity, and sustainability.

To realize the full potential of energy abundance, engagement and responsibility at every level of society are indispensable. This vision requires not only technological innovation but also a deeply rooted commitment from scientists, policymakers, business leaders, and everyday citizens. Each group plays an essential role in cultivating a future where energy is no longer a privilege but a universal asset, where power is both renewable and restorative, fueling human potential and environmental recovery.

For scientists, this vision represents a call to push the boundaries of knowledge, exploring the realms of zero-gravity superconductivity, advanced materials, and efficient transmission methods that make self-sustaining energy systems possible. Scientific research is the backbone of this transformation, with each breakthrough bringing humanity closer to a world where energy can be generated and compounded without compromising environmental integrity. In the labs and workshops where these advancements take shape, scientists hold the keys to unlock a sustainable future, providing the foundation upon which the entire energy-compounding model rests. Beyond technical expertise, scientists must embrace a spirit of stewardship, recognizing that their innovations have the power to shape civilization's trajectory and impact generations to come.

Policymakers and government leaders hold an equally vital responsibility, bridging scientific advancements with real-world applications. They have the authority to craft policies, set regulations, and prioritize public resources that bring the vision of energy abundance into reality. By creating a supportive framework, governments can facilitate the development and deployment of energy-compounding systems, investing in the infrastructure needed to transmit and distribute power equitably. Policymakers must align their agendas with the goals of sustainability, inclusivity, and long-term resilience, ensuring that the transition to energy abundance benefits not only the present but future generations as well. It is within their power to establish the standards that prevent monopolization, protect public access, and foster international cooperation—a cooperative framework that prioritizes the welfare of all people over isolated interests.

Industries and businesses, too, have a critical part to play. In an economy driven by energy abundance, companies must redefine their roles as stewards of both technology and the environment. The shift away from fossil fuels toward a regenerative energy infrastructure calls on corporations to innovate sustainably, aligning profit with the principles of social and ecological responsibility. By investing in renewable technologies, sustainable manufacturing, and eco-friendly practices, industries can position themselves as leaders in a new era of ethical capitalism, one that values long-term resilience over short-term gains. Corporate commitments to transparency, fair distribution, and resource conservation strengthen the energy-compounding model, demonstrating how business interests can coexist with the well-being of the planet. As industries integrate these values into their operations, they help create an economy where sustainability is not an optional virtue but a foundational element of success.

Citizens, as the beneficiaries and stakeholders of this energy transition, wield a unique influence. Public support, informed engagement, and active participation are essential to drive this transformation forward. Individuals play a powerful role in holding leaders accountable, advocating for sustainable policies, and embracing energy-efficient practices in their daily lives. In this way, citizens are not only consumers of energy but custodians of the systems that produce it. Their commitment to conservation, awareness, and responsible energy use is integral to sustaining the abundance model, fostering a cultural shift that views energy as a shared resource rather than a commodity to exploit. When communities rally around the vision of energy abundance, they become active participants in a societal paradigm shift, promoting values that prioritize ecological balance, social equity, and collective prosperity.

To secure this future, humanity must lay a robust ethical foundation. The shift to energy abundance brings with it unprecedented power, and with that power comes an equally profound responsibility. The availability of boundless energy could either empower societies to achieve sustainable harmony or lead to new forms of imbalance and control if mismanaged. It is crucial to approach this transformation with a framework that promotes fairness, protects the rights of all communities, and prevents the concentration of energy resources in the hands of a few. The principles guiding this energy transition must be rooted in transparency, inclusivity, and accountability, ensuring that energy remains a public good and not a private asset.

The establishment of governance structures that oversee this transition is paramount. A system of checks and balances, designed to monitor the ethical use of energy-compounding technologies, can protect against potential abuses and safeguard public interests. Governments, in collaboration with international organizations, must develop policies that promote equitable distribution, discourage exploitation, and empower communities to participate in decision-making processes. This governance should reflect the diverse voices of society, incorporating perspectives from various regions, cultures, and socioeconomic backgrounds to create a system that serves all humanity. A global council or coalition dedicated to energy equity and environmental integrity could provide oversight, ensuring that no single entity or group can disproportionately influence or control the flow of energy.

Moreover, this ethical foundation must emphasize the protection and restoration of the environment. The transition to energy abundance offers humanity a rare opportunity to shift from a model of extraction to one of regeneration. By prioritizing ecological stewardship, society can reverse the environmental damage caused by centuries of fossil fuel dependence, restoring ecosystems and safeguarding biodiversity. Large-scale reforestation, habitat preservation, and climate remediation projects would become feasible with an abundant energy supply, transforming humanity from a contributor to environmental degradation to a champion of ecological renewal. In this model, energy is not just a force that powers human civilization but a tool for healing and harmony with nature.

As humanity enters this new energy era, the potential for cultural transformation is profound. Energy abundance invites society to reconsider its relationship with progress, reframing success not as a measure of consumption but as a balance between innovation and sustainability. In a world where energy scarcity is no longer a constraint, the focus shifts from competition to cooperation, from accumulation to restoration. This new cultural ethos fosters a mindset that values long-term resilience, ecological responsibility, and a shared commitment to a sustainable future.

In this cultural shift, education plays a pivotal role, shaping the mindset of future generations who will inherit the systems and values created today. Schools, universities, and public institutions have a responsibility to instill principles of sustainability, ethical responsibility, and environmental stewardship. By teaching young people about the principles of energy compounding, ecological balance, and shared resources, education empowers them to become informed and engaged custodians of the future. This commitment to knowledge-sharing and ethical awareness lays the groundwork for a society that values harmony with nature as a core tenet of progress, creating a legacy of sustainability that reaches beyond the confines of individual lifetimes.

The transition to an energy-abundant future also provides an opportunity to reimagine economic structures. Traditional models, often driven by scarcity, have emphasized competition and growth at the expense of the environment and social equity. In contrast, an energy-compounding economy is based on principles of regeneration, inclusivity, and resilience. This model supports an economy where resources are not exhausted but reinvested, creating a cycle of growth that sustains itself without depleting the planet's finite reserves. In this economy, industries that prioritize sustainability, energy efficiency, and community well-being are rewarded, encouraging a shift from extractive practices to regenerative ones. By integrating these principles into the fabric of global commerce, society can build an economy that thrives within the limits of the planet while elevating the quality of life for all.

The vision of energy abundance challenges humanity to imagine a world where power is no longer a divisive force but a unifying one, where energy serves as the foundation of a just and sustainable society. It offers a roadmap for addressing the interconnected crises of climate change, poverty, and inequality, empowering societies to transcend the limitations of scarcity. In a world where energy flows freely and sustainably, the potential for human achievement expands, unlocking new horizons in science, healthcare, education, and technology. This vision is not just a blueprint for energy infrastructure; it is a framework for reimagining the very essence of civilization.

As humanity stands on the brink of this new era, it is invited to dream of a future that honors both human ingenuity and the planet's natural balance. The journey to energy abundance is one of shared ambition, responsibility, and opportunity—a collective endeavor that transcends national borders, cultural divides, and historical inequalities. By embracing the principles of energy compounding, society can create a world where power fuels not only technological advancement but also ethical and environmental progress. This legacy of energy abundance is one of harmony, where human achievement is measured not by consumption but by its capacity to uplift, renew, and sustain.

In moving toward this vision, each step reinforces a profound commitment to shared progress. The energy-compounding model encourages society to think beyond the immediate and embrace a perspective that values long-term harmony, recognizing that true progress is rooted in balance. It challenges humanity to envision an inclusive future where all communities benefit from the fruits of innovation and where environmental preservation is not an afterthought but a primary objective. By honoring these values, the journey to energy abundance becomes more than a technological achievement; it becomes a testament to the human spirit's capacity for resilience, creativity, and unity.

The call to action is clear: to achieve energy abundance, humanity must embrace a new paradigm that prioritizes ethical responsibility, global collaboration, and sustainable progress. This vision is not a solitary pursuit but a collective aspiration that binds generations in a shared mission. As scientists, leaders, industries, and citizens come together to realize this future, they build a world where energy scarcity is no longer a barrier, where the Earth and its people thrive in unison. The journey to energy abundance is, ultimately, a journey toward a more compassionate, sustainable, and just world—a legacy that will shape the course of human history for centuries to come.

As humanity embarks on the path to energy abundance, it steps into a future defined by the values and principles that ensure this achievement serves the common good. Energy compounding, as a vision, is more than a technological triumph; it is a symbol of human ingenuity and an embodiment of our shared commitment to balance, inclusivity, and resilience. With this advancement, humanity can transition from an age characterized by scarcity and conflict over resources to one of restoration, opportunity, and stewardship, leaving behind a legacy that future generations will inherit with pride and purpose.

In this energy-abundant future, the successes of science and innovation are coupled with profound shifts in societal structures and individual perspectives. As energy becomes accessible and reliable for all, communities can flourish without compromising environmental integrity or exhausting finite resources. The landscapes of cities and rural regions alike are transformed, freed from pollution and empowered by clean, sustainable energy sources. No longer reliant on fossil fuels, societies invest in infrastructure that integrates seamlessly with natural ecosystems, from green architecture that minimizes environmental impact to urban farming systems that ensure food security. By transcending the traditional divides of wealth, location, and access, this new infrastructure creates a harmonious interplay between human life and the environment, embodying the principles of responsible stewardship and respect for nature's rhythms.

The cultural impact of energy abundance is equally transformative, as the relentless pursuit of resources is replaced by a collective ethos of stewardship and renewal. In this world, success is measured not by the scale of consumption but by the extent of restoration and creativity. Artists, scientists, and visionaries collaborate freely, unimpeded by the limitations once imposed by resource scarcity. The compounding effect of shared knowledge and innovation accelerates progress, fostering a global culture of collaboration that values the well-being of both people and the planet. Schools, libraries, and public institutions become centers of shared learning and ethical inquiry, nurturing a generation that understands energy not just as a tool but as a shared resource that carries an inherent responsibility. In this world, energy abundance is celebrated not merely for what it enables but for how it enriches lives and strengthens communities.

As humanity's collective focus shifts toward sustainability and equity, the very structure of economies transforms. An energy-compounding economy does not adhere to the extractive, competition-based frameworks of the past. Instead, it aligns growth with environmental stewardship, rewarding practices that conserve resources and uplift communities. Corporations and industries that once relied on the depletion of natural resources pivot toward regenerative practices, from sustainable agriculture and closed-loop manufacturing to recycling and zero-waste production. Energy-intensive sectors—such as construction, transportation, and manufacturing—undergo fundamental shifts, incorporating eco-friendly technologies that reduce carbon footprints and enhance the longevity of both products and environments. In this economy, innovation is directed toward the preservation and optimization of resources, ensuring that growth and prosperity benefit not only the present generation but also those to come.

Beyond economic transformation, energy abundance empowers global stability and peace. Many of the world's conflicts have stemmed from the struggle over finite resources, with nations vying for control over oil, minerals, and other essential commodities. In a world where energy is readily available and sustainably produced, such conflicts diminish, replaced by opportunities for cooperation and shared progress. Nations, no longer constrained by energy dependencies, can reallocate resources toward education, healthcare, and social welfare, creating resilient societies that prioritize the well-being of their people. Energy abundance thus becomes a force for peace, fostering diplomatic collaboration and mutual respect. In this new geopolitical landscape, alliances are built on shared environmental goals and commitments to ethical progress, enabling humanity to address global challenges as a united force.

The environmental impact of this vision is profound. With an abundance of clean, renewable energy, humanity can undertake large-scale projects that restore and protect ecosystems. Reforestation initiatives, powered by sustainable energy, can revitalize areas affected by deforestation, restoring biodiversity and contributing to climate stabilization. Conservation efforts benefit from the support of technologies that monitor and protect endangered habitats, ensuring that natural ecosystems thrive alongside human development. Clean energy also enables the deployment of advanced environmental protection technologies, from ocean cleanup projects to atmospheric carbon capture, directly addressing the impacts of climate change. By embracing energy abundance, humanity gains the tools to not only sustain the planet but to heal it, transforming Earth from a resource to exploit into a home to cherish and protect.

In this vision of the future, energy abundance does not simply alleviate the challenges of today; it opens doors to a future where human potential knows no bounds. With sustainable energy as a foundation, humanity can pursue the boldest dreams of exploration, knowledge, and growth. Scientific research, unburdened by energy constraints, can delve deeper into the mysteries of the universe, from quantum physics to deep space exploration. With abundant power, humanity could journey to the stars, exploring the cosmos with a sense of unity and purpose that reflects the spirit of shared stewardship cultivated on Earth. Energy abundance thus becomes the bedrock of human evolution, enabling achievements that transcend current limitations and inspiring generations to come.

The cultural and philosophical legacy of energy abundance is one of compassion, responsibility, and wisdom. This era invites humanity to redefine what it means to be a custodian of the planet, encouraging societies to view progress through the lens of ethical and environmental impact. Energy abundance challenges us to prioritize the common good, fostering a global culture that values both human dignity and ecological balance. As humanity's relationship with energy evolves, so too does its understanding of itself, recognizing that true power lies not in dominance but in the ability to uplift, protect, and renew.

In moving toward this future, every action, every policy, and every technological advancement contributes to a legacy that reflects the highest aspirations of humanity. Energy abundance is not merely an end goal but an ongoing commitment—a journey that calls for vigilance, cooperation, and continuous ethical reflection. As humanity builds this energy-compounding infrastructure, it also constructs a cultural and moral foundation that will guide future generations. In doing so, we leave behind a world where energy serves not as a source of division or destruction but as a catalyst for unity and creation.

This vision of energy abundance embodies the ideals of sustainability, equity, and shared progress. It is a legacy that honors the past by learning from it, respects the present by addressing its needs, and celebrates the future by making it possible. In this era, humanity finds a profound sense of purpose, embracing the role of stewards of the Earth and architects of a sustainable civilization. Energy, once a source of limitation, now illuminates a path forward, guiding societies toward a destiny shaped not by scarcity but by the boundless possibilities of renewal, compassion, and shared endeavor.

As humanity stands on the cusp of this transformative journey, it is invited to look beyond immediate challenges and envision a world where energy abundance is a beacon of hope and harmony. This vision, driven by the principles of energy compounding, sets forth a future where technology and ethics converge, where human potential flourishes without exhausting the planet's resources, and where prosperity is shared, sustainable, and resilient. By embracing this future, humanity crafts a legacy that extends beyond generations—a testament to our shared commitment to build a world where energy powers not just our machines but our highest ideals, lifting civilization to new heights of wisdom, harmony, and collective strength.

Prologue: The Energy Challenge of the 21st Century

As humanity stands at the threshold of the twenty-first century, it finds itself immersed in a profound energy crisis. This crisis is not only a consequence of rising demand but is deeply intertwined with the ecological and socio-political fabric of modern society. The surge in global population, the rapid urbanization of once-rural landscapes, and the expansion of energy-intensive industries have all placed unprecedented stress on our energy infrastructure. From industry to agriculture, from healthcare to transportation, every aspect of life relies on energy as its backbone. As billions strive for a higher quality of life and as developing regions aim to industrialize, the demand for energy shows no signs of slowing down. This relentless growth in demand has brought the world to a critical juncture, one where traditional methods of energy production struggle to keep pace, and innovative solutions are not merely a luxury but a necessity.

Conventional energy systems, largely based on fossil fuels, are inherently limited. They are finite, bound by the natural reserves that exist within the Earth, and are accompanied by significant environmental and health impacts. Fossil fuels, though foundational in propelling modern society, are double-edged; they enable progress but at the cost of environmental degradation and resource depletion. Renewable energy sources like solar, wind, and hydroelectric power have emerged as alternatives, but they too face challenges of intermittency, storage limitations, and geographic dependency. Thus, the question of energy abundance is no longer just a matter of fulfilling rising demand but is essential for achieving sustainable development, social equity, and long-term environmental resilience. Without a transformative shift toward abundant and clean energy, humanity risks aggravating climate change, fostering geopolitical conflicts over dwindling resources, and exacerbating economic disparities that arise from unequal access to power.

The pressures exerted by this rising demand present a daunting challenge, yet they also offer a remarkable opportunity. The need for energy abundance acts as a catalyst, encouraging society to consider bold, innovative approaches that transcend the limitations of current systems. In an age when technology can make the improbable seem feasible, the prospect of energy compounding and self-sustaining power systems heralds a new frontier. These developments allow us to imagine a world where energy is not confined by scarcity but is an ever-present, self-renewing resource that supports both human innovation and environmental restoration. As demand continues to rise, this new paradigm offers a solution that could reverse the cycle of scarcity, fostering a future where humanity's progress is harmonized with the planet's ecological health.

The environmental impacts of current energy sources are as profound as they are far-reaching, influencing ecosystems, human health, and global climate stability. Fossil fuels, still dominant worldwide, emit massive quantities of carbon dioxide and other greenhouse gases, directly contributing to global warming and climate change. The extraction of coal, oil, and natural gas entails environmental disruption, often leading to deforestation, water contamination, and soil erosion. These consequences ripple through ecosystems, pushing numerous species toward endangerment and extinction and destabilizing natural cycles essential for a balanced environment.

Even renewable energy sources, although cleaner in operation, are not free from ecological impact. Solar and wind installations require extensive land, while the production and disposal of solar panels and batteries involve toxic chemicals and rare materials that pose environmental risks. Hydropower, although renewable, disrupts river ecosystems, altering water quality and flow patterns critical to aquatic life. In this context, the need for an energy revolution becomes undeniable. Humanity must transcend reliance on finite resources and the environmental costs tied to renewables, seeking a way to generate power that does not merely shift one form of environmental strain to another. The vision of self-sustaining, zero-gravity energy systems and innovative energy compounding methods emerges as a pathway to provide power without the ecological toll, offering a sustainable energy source that meets growing needs while preserving the planet's health.

The implications of energy abundance extend beyond environmental considerations, reshaping the political and economic landscapes that underpin modern society. An energy system free from scarcity promises to alter not only the mechanics of production and distribution but also the very nature of energy as a political force. Energy scarcity has long driven competition between nations, as they vie for access to finite resources necessary for stability and growth. This scramble for fossil fuel reserves has historically influenced alliances, sparked conflicts, and defined the contours of global power. The advent of energy-compounding technology disrupts this paradigm, enabling a shift where energy is no longer a limited commodity but a renewable, universally available resource.

For countries dependent on energy exports, this shift poses a challenge to long-standing economic models. Nations that have built their economies on fossil fuel revenues may need to undergo extensive restructuring as global reliance on these resources diminishes. This economic transition will impact government revenue and employment, necessitating policy frameworks that support economic diversification and workforce adaptation. Conversely, for energy-importing nations, energy abundance represents a chance for greater self-sufficiency, reducing dependence on foreign suppliers and fostering stable economic growth. The implications of this balance reach deep into the international arena, fostering a cooperative approach over resources that have historically divided.

From an economic perspective, an abundant energy system capable of self-sustaining generation holds the potential to lower production and distribution costs across all sectors. Reliable, affordable energy could drive efficiencies across industries from manufacturing to agriculture, eliminating traditional cost barriers and spurring innovation. Small businesses, especially in developing economies, would gain competitive footing, promoting economic inclusion and reducing global disparities. As energy abundance becomes a reality, it could empower nations to invest in social services, infrastructure, and sustainable development without the financial constraints associated with energy scarcity.

These political and economic transformations underscore the importance of responsible governance. The benefits of energy abundance are vast, yet they must be managed with a commitment to transparency, equity, and access for all. The monopolization of such a resource would undermine its potential as a global equalizer. Ensuring that energy abundance serves as a universal asset, free from exploitation or exclusive control, requires collaborative governance frameworks that prioritize public welfare and ethical innovation. These considerations affirm that the journey toward energy abundance is not just a technological evolution but a societal one, one that calls for a balanced, ethical approach to harnessing power for collective prosperity.

At the core of this transition lies the transformative role of technology. As existing energy sources strain under environmental and resource-based limitations, advanced technology emerges as the key to unlocking a sustainable and abundant energy future. Technology offers not only the tools to mitigate climate change but the foundation to rethink energy from the ground up. Concepts such as energy compounding, superconductivity, and space-based power systems hold the potential to redefine infrastructure, paving the way for energy models that are self-renewing and scalable beyond the constraints of Earth's resources.

A central feature of this technological renaissance is the capacity to produce surplus energy, shifting from the linear dynamics of traditional power systems to regenerative models. Advanced technology enables each unit of generated energy to sustain itself while producing excess output, forming a compounding loop that aligns with the financial principles of reinvestment. In this self-sustaining system, energy becomes an expanding resource, creating a feedback loop that supports ever-increasing output without the need for additional inputs. This shift marks a departure from conventional models, where energy output is tightly coupled to input, and heralds a new era where energy amplifies in harmony with societal needs.

The advancement of technology in energy production is not only a practical response to dwindling resources; it is a visionary endeavor that could revolutionize how humanity relates to power and progress. Through the application of zero-gravity energy generation, superconductivity, and innovative transmission methods, the concept of energy compounding pushes the boundaries of what energy can achieve. At its essence, energy compounding represents a shift from mere consumption to a model that reinvests its output back into the system. This regenerative cycle parallels the principles of financial compounding, where small, continuous gains accumulate over time to create exponential growth. In this context, energy becomes a living force, amplifying itself with each cycle rather than depleting finite resources.

Zero-gravity conditions offer a unique environment for realizing this compounding model. In the weightless realm of space, traditional constraints like friction and gravitational pull are removed, allowing machines to operate with greater efficiency and minimal energy loss. The frictionless setting of space preserves the structural integrity of materials and systems, meaning that motors and generators experience less wear and tear, enabling them to maintain a constant state of operation with fewer interruptions. This remarkable efficiency, paired with the potential of superconductors, paints a vision of energy systems that work continuously, producing and reinvesting surplus energy without succumbing to the degradations that define Earth-bound systems.

The integration of superconductivity into energy compounding systems adds another layer of revolutionary potential. Superconductors are materials that, when cooled to ultra-low temperatures, conduct electricity without resistance. On Earth, energy flowing through a conductor inevitably meets resistance, transforming a portion of it into heat and leading to energy loss. Superconductors eliminate this issue, allowing energy to circulate through systems indefinitely without leakage. When applied in space, where the naturally low temperatures support superconductivity, these materials enable near-perfect energy retention. The resulting system has the capability to achieve a level of operational efficiency that surpasses even the most advanced terrestrial technologies. By removing resistance from the equation, superconductors allow each generated unit of energy to be preserved, compounding the initial input and accelerating growth without external resource depletion.

The implications of this system extend beyond the theoretical; they hint at a practical future where energy compounding reshapes the foundational infrastructure of global energy production and distribution. With strategically placed energy-compounding stations in space, located at stable points known as Lagrange points where gravitational forces balance, energy can be generated continuously and then transmitted to Earth. These stations would operate with unparalleled stability, using minimal resources to remain in orbit while benefiting from constant access to solar energy. Positioned in these optimal locations, they would receive uninterrupted sunlight, amplifying their energy input and supporting a system that not only sustains itself but generates an excess that can be distributed back to Earth.

In tandem with these advancements, the evolution of wireless energy transmission will play a pivotal role. Nikola Tesla, the brilliant inventor and visionary, once imagined a world where energy could be transmitted without wires, made accessible to all, unhindered by distance or physical barriers. Although his early experiments were constrained by the technological limits of his time, his pioneering ideas laid the groundwork for modern efforts in resonant inductive coupling and space-based energy relay systems. By harnessing wireless transmission, these space-based energy stations could beam power directly to Earth, revolutionizing how energy reaches communities, especially those that are underserved by current infrastructure. The convergence of these technologies—superconductivity, zero-gravity operation, and wireless transmission—represents the birth of a system that is as efficient as it is revolutionary.

This vision of a sustainable, compounding energy system calls for humanity to think beyond the conventional limitations of scarcity, infrastructure, and geography. With zero-gravity energy generation, humanity can begin to conceive of power not as a finite resource to ration but as a boundless wellspring that grows with use. Just as natural ecosystems thrive through cycles of regeneration, an energy-compounding system enables power to flow in a continuous loop, sustaining itself and expanding as it meets increasing demand. This abundance empowers communities and economies previously limited by energy scarcity to grow, innovate, and thrive.

As we contemplate the implementation of such a system, it becomes clear that energy abundance could reshape society on a structural level. The absence of energy scarcity would redefine the way industries operate, removing limitations on production and transportation, reducing costs, and promoting innovation. Entire industries could grow sustainably, free from the constraints imposed by traditional energy economics. Sectors like manufacturing, agriculture, and transportation could function at unprecedented levels of efficiency, as the costs associated with energy are minimized. The implications for healthcare, education, and technology are equally profound, as reliable and inexpensive energy becomes the foundation upon which modern civilization builds its next advancements.

For individuals and communities, energy abundance translates to opportunity. With continuous, reliable access to power, remote regions and developing areas could flourish, achieving levels of growth that were once unimaginable. The availability of energy in every corner of the globe would enable advancements in education, healthcare, and industry, elevating the standard of living and empowering communities to pursue sustainable development. For the first time in history, energy becomes not a privilege of geography or economic status but a universally available resource, fostering equity and lifting millions out of poverty. In an energy-abundant world, quality of life is no longer tethered to resource constraints, and societies can focus on higher pursuits, from cultural enrichment to environmental stewardship.

The environmental impact of transitioning to a self-sustaining, space-based energy system is transformative. Fossil fuel dependence would gradually diminish, reducing emissions and slowing climate change. Vast areas of land, once dominated by coal mines, oil fields, and gas pipelines, could be reclaimed, allowing ecosystems to heal and biodiversity to flourish. Forests, rivers, and oceans would benefit from the reduction in pollution, and species threatened by habitat loss and climate disruption would have a chance to recover. Clean energy abundance creates a world where human progress no longer comes at the cost of environmental degradation but instead supports a balanced relationship with nature. This paradigm aligns human aspirations with ecological health, creating a legacy of harmony rather than exploitation.

Moreover, an energy-abundant future offers solutions to pressing environmental issues that were previously limited by resource constraints. With limitless, clean energy, society could undertake ambitious environmental restoration projects, from large-scale reforestation and ocean cleanup efforts to carbon capture initiatives that remove greenhouse gases from the atmosphere. Abundant power would support advanced water purification technologies, making clean water accessible even in arid regions, and would enable sustainable agriculture practices that reduce dependency on finite natural resources. In this transformed world, energy abundance serves as a catalyst for environmental renewal, enabling humanity to become a steward of the Earth rather than a burden on it.

In reimagining energy as an unlimited, universally accessible resource, humanity must also confront the ethical responsibilities inherent in such power. Energy abundance, while offering vast potential for growth, must be managed with a commitment to equity and responsibility. History has shown that resources, when monopolized, can lead to disparities and exploitation. The governance of an energy-compounding infrastructure must prioritize transparency, ensuring that access to energy remains a public good rather than a tool of control. This ethical commitment requires cooperation between nations, industries, and institutions, creating regulatory frameworks that protect the integrity of energy abundance as a shared asset. This system calls for inclusive policies, equitable access, and collaborative oversight to ensure that the benefits of energy compounding reach every individual, regardless of geography or socioeconomic status.

The establishment of global oversight institutions dedicated to managing energy abundance could foster cooperation and accountability, allowing humanity to navigate the challenges of energy distribution with foresight and integrity. These institutions would act as custodians, protecting against the monopolization of energy resources and advocating for fair practices that promote public access. By managing energy as a shared resource, humanity builds a foundation of trust, ensuring that the fruits of technological innovation are accessible to all. This cooperative approach establishes energy abundance as a force for unity rather than division, transforming it into a tool for collective progress and stability.

Education and public awareness play pivotal roles in guiding society through this energy transition. An energy-abundant future requires an informed citizenry that understands the principles of sustainability, conservation, and technological stewardship. Schools, universities, and public institutions must educate future generations about the ethical responsibilities and environmental imperatives associated with energy abundance. By fostering a culture of awareness and responsibility, society empowers individuals to contribute positively to the global shift toward sustainable energy. Knowledge-sharing becomes a central value, promoting collaboration and transparency in managing the abundant resources that define this new era.

In this vision, energy abundance does more than alleviate scarcity; it transforms civilization's relationship with power, resource management, and social equity. This new paradigm inspires societies to measure success not by consumption but by sustainability, fostering a global culture that values balance, regeneration, and environmental resilience. With energy no longer a limiting factor, human potential can be realized in ways previously restricted by resource scarcity. Scientific research, artistic pursuits, and cultural initiatives are empowered to thrive in an environment where the constraints of energy dependency are replaced by an ethos of sustainable innovation and shared growth.

The shift toward energy abundance invites a cultural renaissance, reorienting values around environmental stewardship, equity, and human progress. As power flows freely and sustainably, societies can invest in endeavors that enrich life and build resilience. Infrastructure can be designed not merely for utility but for harmony with the environment, creating cities and communities that coexist with nature. As the need for finite resource extraction diminishes, humanity can focus on projects that restore rather than exploit, healing the planet and establishing a legacy of regeneration.

In an energy-abundant world, society undergoes a profound transformation in values, where success is no longer measured by the accumulation of resources but by the pursuit of sustainability, equity, and harmony. Freed from the constraints of energy scarcity, humanity has the unprecedented opportunity to redefine its collective purpose and focus on a vision of progress that benefits not only itself but also the ecosystems with which it shares the planet. This shift toward a regenerative economy, built on the principles of renewable energy and environmental stewardship, promises a future where human activity and natural cycles are aligned, creating a world that nurtures growth without exploitation. The result is a civilization that embraces the notion of "enough"—not as a limitation but as a powerful guiding principle, where prosperity is defined by balance rather than excess.

Cultural values evolve to reflect this new relationship with energy. When humanity no longer needs to compete for finite resources, cooperation and collective well-being naturally become core societal goals. Communities previously divided by energy access disparities are empowered to collaborate, fostering a spirit of unity that transcends borders. In a world where energy abundance is a shared reality, societies can reimagine economic structures and social norms, prioritizing the distribution of resources based on need and sustainability rather than scarcity and wealth. This paradigm creates a cultural shift toward shared prosperity, where the pursuit of knowledge, creative expression, and mutual support become the driving forces of human development.

At the heart of this transformation is the realization that abundance brings responsibility. Access to limitless, clean energy imposes ethical obligations on both individuals and societies, urging them to consider the broader impacts of their actions on the planet and future generations. This responsibility fosters a culture of mindfulness, where energy consumption is approached with respect and consideration. Rather than taking abundance for granted, societies instill values of conservation, awareness, and appreciation for the resources they possess. Schools, community organizations, and public institutions work collectively to promote environmental literacy and ethical stewardship, teaching future generations that energy, even in abundance, is not something to waste. Through this cultural shift, humanity learns to view resources as shared assets, reinforcing a sense of accountability and interconnectedness that underpins sustainable progress.

The shift toward sustainable energy also invites a reevaluation of how communities are structured, designed, and maintained. Cities and towns evolve to incorporate renewable energy sources seamlessly into everyday life, creating urban landscapes that operate in harmony with nature. Smart cities, powered by a constant flow of clean energy, become hubs of innovation and efficiency. Buildings are designed with integrated solar panels, green roofs, and energy-efficient systems, reducing environmental impact and enhancing the quality of life for residents. Public transportation networks, driven by abundant, renewable power, replace reliance on fossil-fuel-based vehicles, reducing pollution and fostering healthier, more sustainable cities. In rural areas, energy abundance enables decentralized power grids that support agricultural advancements, water purification, and sustainable resource management, creating resilient communities capable of thriving without dependence on distant infrastructures.

Energy abundance also transforms education, empowering individuals to pursue knowledge without the limitations imposed by energy scarcity. Schools, colleges, and research institutions become centers of exploration, where innovation and creativity flourish unburdened by financial and resource constraints. Students have the freedom to learn, explore, and contribute to society without the fear that their aspirations will be hindered by energy access limitations. Educational institutions can operate at full capacity, leveraging abundant energy to power advanced technologies, laboratories, and digital learning platforms that provide equal opportunities to all students, regardless of socioeconomic background. This commitment to equitable education fosters a generation of environmentally conscious, ethically aware citizens who are prepared to lead society into a future defined by responsible abundance.

Healthcare, too, is revolutionized in an energy-abundant world. With access to consistent, renewable energy, medical facilities no longer face power shortages, allowing healthcare providers to deliver uninterrupted, high-quality care to every patient. Rural and underserved communities benefit as hospitals and clinics, empowered by energy compounding systems, can operate advanced equipment, maintain optimal environmental controls, and provide continuous care without the fear of energy outages. The abundance of clean energy enables the use of advanced medical technologies, from MRI machines to telemedicine networks, that connect patients with specialists regardless of geographic barriers. With this reliable infrastructure in place, healthcare becomes a universal right, available to all people regardless of location, and societies are healthier, more resilient, and better equipped to face public health challenges.

The economic impact of energy abundance ripples through every industry, from manufacturing to information technology. Production processes evolve to harness renewable energy sources, resulting in lower operational costs, increased efficiency, and a reduction in environmental impact. Factories and supply chains operate sustainably, producing goods that are no longer tied to the finite availability of traditional resources. This transition encourages companies to adopt regenerative practices, manufacturing products designed for longevity, recyclability, and minimal environmental footprint. This new approach fosters an economy where consumption is no longer wasteful, but mindful and purposeful, transforming industries that have historically relied on resource extraction into sectors focused on environmental restoration and sustainable growth.

In the digital age, information technology thrives in an energy-abundant society. Data centers, which once consumed vast amounts of energy, are now powered sustainably, reducing their environmental impact while supporting the continued expansion of cloud computing, artificial intelligence, and data-driven innovation. Renewable energy supports the exponential growth of technology, enabling advancements in artificial intelligence, machine learning, and quantum computing that were previously constrained by energy limitations. As data processing becomes more efficient, the possibilities for digital innovation expand, empowering researchers and developers to push the boundaries of knowledge, tackle complex problems, and create solutions that benefit society as a whole. The digital infrastructure of an energy-abundant world becomes the backbone of global connectivity, allowing people to collaborate, share ideas, and contribute to a collective pool of knowledge that drives progress on a planetary scale.

Energy abundance also fundamentally reshapes global geopolitics, as access to energy ceases to be a source of conflict. Historically, nations have competed for control over fossil fuels, leading to political tensions, economic dependencies, and conflicts over territorial claims. In an energy-abundant world, this paradigm shifts, replacing competition with cooperation. As all nations gain access to renewable power, global alliances are forged not by necessity but by shared values and common goals. The potential for conflict over energy resources diminishes, paving the way for a more peaceful, stable international community where collaboration and mutual respect replace rivalries rooted in scarcity. This new geopolitical landscape enables nations to pool resources, share technological advancements, and work collectively to address challenges that transcend borders, from climate change and public health crises to economic inequality and environmental degradation.

Governance in an energy-abundant world takes on a renewed sense of purpose and responsibility. Leaders are called upon to prioritize transparency, fairness, and the public good, managing resources as stewards rather than owners. International bodies collaborate to ensure that energy abundance serves all people, establishing guidelines that prevent monopolization and protect against misuse. By upholding principles of accountability, inclusivity, and sustainability, global institutions help create a framework for energy abundance that empowers citizens while safeguarding against potential abuses. Governments and regulatory bodies work together to establish ethical standards, promote equitable access, and reinforce the notion that energy is a universal right, available to all people and protected for future generations.

The environmental impact of energy abundance continues to expand, transcending the elimination of fossil fuel reliance to enable active ecological restoration. With an abundant supply of clean power, humanity can tackle environmental challenges that once seemed insurmountable. Reforestation efforts, backed by renewable energy, become vast and ambitious, restoring degraded landscapes, reviving biodiversity, and stabilizing regional climates. Polluted waterways are purified, with abundant energy supporting advanced filtration systems that remove contaminants and restore ecosystems. Desertified areas can be revitalized through sustainable irrigation practices powered by renewable energy, transforming once-barren regions into fertile land that supports agriculture and biodiversity. Humanity, no longer a threat to the planet's natural balance, becomes its caretaker, using technology and innovation to repair past damage and foster resilience.

In an energy-abundant world, conservation and growth are no longer mutually exclusive. Communities can expand and develop without infringing on natural habitats, as energy supports technologies that minimize ecological impact. Vertical farming, powered by renewable energy, reduces the need for land-intensive agriculture, allowing forests and grasslands to thrive. Marine habitats, once compromised by fossil fuel extraction and pollution, regain health as offshore renewable energy sources replace invasive drilling practices. By embracing a model of regenerative growth, humanity finds a way to prosper while preserving and enhancing the natural world.

With energy abundance, humanity is invited to contemplate a new era of exploration and innovation, where power serves as a tool for discovery rather than destruction. Space exploration, driven by renewable energy sources, expands beyond our planet, allowing humanity to venture further into the cosmos. The availability of sustainable power enables the construction of space habitats, interplanetary stations, and research outposts that deepen our understanding of the universe. Freed from the energy limitations that once bound us to Earth, humanity's spirit of exploration and curiosity thrives, marking a new age of discovery where the pursuit of knowledge knows no boundaries.

In contemplating this energy-abundant future, society must also reckon with the profound ethical considerations that accompany such power. The shift toward abundant energy presents both opportunity and responsibility, compelling humanity to reflect on its role as a steward of the planet and an advocate for equity.

This ethical framework emphasizes the need to protect and share energy as a universal good, ensuring that it serves all people equally. In this vision, power is not a means of control but a force for liberation, uplifting individuals and communities alike. Humanity is called upon to foster a culture of integrity, one that values restraint, respect, and mindfulness in its approach to energy use, ensuring that abundance remains a sustainable gift rather than a catalyst for excess.

The legacy of an energy-abundant world is ultimately one of hope, progress, and unity. It is a future where human potential is not limited by finite resources but empowered by the boundless possibilities of renewable power. In this world, civilization thrives without exploitation, creating a sustainable relationship with the planet and with each other. The achievements of this new era serve as a testament to humanity's capacity for innovation, compassion, and wisdom, offering a vision of progress that honors both the Earth and the generations yet to come. This legacy reflects humanity's highest aspirations, creating a world where energy abundance fuels not only technological advancement but also a shared journey toward peace, sustainability, and collective prosperity.

As humanity steps into a future of energy abundance, the guiding principles of society are inevitably reshaped, inviting deeper reflection on what it means to live sustainably and ethically. Energy abundance challenges humanity to transcend the traditional, scarcity-driven mindsets that have historically guided decisions, economies, and social structures. In a world where power is no longer a limited resource, where energy flows freely and renewably, there is an invitation to redefine prosperity itself—not as the unbridled accumulation of resources but as the balanced and equitable distribution of them. This shift demands not only a recalibration of how society operates but also an introspective look at the values that shape human progress.

At the heart of this new paradigm is the principle of stewardship, a concept that calls for the responsible, mindful management of resources with a view toward sustainability, equity, and long-term resilience. An energy-abundant future demands that humanity exercise restraint and care in wielding this new power, recognizing that the availability of limitless energy does not absolve society from the responsibility to use it wisely. Energy abundance, if mismanaged or monopolized, could become a source of disparity rather than unity, reinforcing old patterns of inequality. Therefore, the governance of energy abundance must be grounded in the principles of fairness, transparency, and accessibility, ensuring that all individuals, communities, and nations benefit from this shared resource.

The ethic of stewardship extends beyond governance to encompass individual actions and societal norms, fostering a culture where energy is valued not for its ubiquity but for its potential to uplift, empower, and transform. This cultural shift fosters a collective commitment to conservation, even in the face of abundance, recognizing that restraint and mindfulness are as critical to sustainability as technological advancement. Educational initiatives, community programs, and public discourse must work together to nurture this ethic, encouraging each person to consider the broader implications of their energy use. Through this cultural evolution, energy abundance becomes not a license for excess but an opportunity to cultivate responsibility, humility, and a deep respect for the interconnectedness of all life on Earth.

The notion of equity becomes a foundational pillar in this vision of energy abundance, emphasizing the need to bridge historical divides and create a world where access to resources is a universal right rather than a privilege. This vision seeks to dismantle the barriers that have long separated regions and communities based on energy availability, empowering even the most remote or underserved populations to participate fully in the global economy. By ensuring that energy abundance is accessible to all, society can address the root causes of many economic and social inequalities, allowing communities to thrive without the limitations imposed by energy scarcity. This equitable access to energy paves the way for a world where human potential is not bound by geography or economic status, enabling each person to contribute to and benefit from collective progress.

For nations that have historically relied on fossil fuel exports as an economic foundation, the transition to energy abundance presents both challenges and opportunities. The shift away from nonrenewable energy sources requires these economies to reimagine their role in a global society that no longer depends on oil, coal, or natural gas. While this transition may initially disrupt established economic structures, it also offers a chance for diversification and innovation. Governments and industries within these countries have the opportunity to invest in renewable technologies, sustainable infrastructure, and regenerative industries that align with the principles of energy abundance. This reimagined economy, built on the pillars of sustainability and resilience, positions former fossil fuel-dependent nations as leaders in the new global paradigm, contributing to a world where prosperity is both shared and enduring.

For energy-importing nations, energy abundance represents the promise of autonomy and self-sufficiency. No longer dependent on external resources, these nations can build resilient energy infrastructures that support stable growth, economic security, and environmental integrity. Freed from the constraints of energy imports, they gain the flexibility to allocate resources toward social services, technological advancement, and infrastructure development, fostering a higher standard of living for their citizens. This newfound self-sufficiency strengthens both domestic and international relations, reducing economic dependencies that have historically fueled conflict and fostering a spirit of cooperation rooted in shared abundance rather than competitive scarcity.

With energy abundance, the role of international cooperation transforms from a necessity driven by scarcity to a choice motivated by shared goals. The universal availability of renewable energy creates the conditions for global partnerships that are proactive and constructive, focusing on mutual advancement rather than defensive resource management.

Nations collaborate to share technological innovations, regulatory frameworks, and best practices that enhance the efficiency, distribution, and ethical governance of energy systems. By working together, countries create a network of knowledge exchange that amplifies the benefits of energy abundance, ensuring that advancements in one region contribute to the well-being of all. This spirit of cooperation fosters a global society where nations stand united in their commitment to sustainability, equity, and collective progress.

The abundance of energy also redefines humanity's relationship with the environment, allowing for a fundamental shift from exploitation to restoration. Freed from the need to extract finite resources, society gains the capacity to focus on healing and protecting ecosystems that have long suffered from industrial expansion and pollution. Energy abundance empowers large-scale environmental restoration projects that once seemed out of reach, transforming humanity from a force of extraction to one of regeneration. Entire landscapes can be revitalized, and ecosystems restored to a state of balance, creating a future where human activity and natural cycles coexist harmoniously. This shift fosters a relationship with nature that is not adversarial but symbiotic, where the advancement of civilization supports, rather than undermines, the health of the planet.

The availability of clean, limitless energy supports projects like reforestation, soil restoration, and ocean cleanup, helping to mitigate the impacts of climate change and restore biodiversity. These initiatives become feasible on an unprecedented scale, driven by the understanding that energy abundance can fuel endeavors that heal the planet as effectively as it once powered industries that depleted it. The energy needed to reclaim arid landscapes, purify contaminated water sources, and protect endangered species is readily available, allowing society to take bold steps in preserving and enhancing the natural world. By investing in these restoration efforts, humanity demonstrates a commitment to leaving a legacy of ecological health and stability, offering future generations a world that is more vibrant, diverse, and resilient.

Energy abundance further supports the emergence of technologies that not only reduce environmental impact but actively contribute to ecological resilience. Carbon capture, for instance, becomes a viable solution for mitigating greenhouse gas emissions, allowing society to reverse some of the damage inflicted by centuries of fossil fuel use. Renewable energy supports the infrastructure required for carbon capture on a global scale, enabling the extraction and storage of atmospheric CO_2 in ways that help stabilize the climate. Advanced water purification systems, powered by abundant energy, make clean water accessible even in regions where natural water sources have been compromised, addressing one of humanity's most pressing environmental and humanitarian challenges.

In this new era, humanity is called to be not only a custodian of energy resources but also an advocate for ethical and sustainable technological progress. The development of renewable energy technologies must align with values that prioritize transparency, equity, and long-term thinking. The path toward energy abundance offers humanity the opportunity to adopt principles of responsible innovation, ensuring that advancements in technology are pursued with foresight, caution, and an unwavering commitment to the greater good. This responsibility extends to industries, research institutions, and policymakers, who must work collaboratively to create frameworks that encourage ethical practices, prevent monopolization, and safeguard the interests of all people.

The influence of energy abundance on society's cultural fabric is equally profound. Freed from the pressures of scarcity, communities can shift their focus from survival to enrichment, fostering a culture of creativity, learning, and mutual respect. The arts, sciences, and humanities flourish as people are empowered to pursue their passions without the constraints of limited resources. Art and innovation become central to public life, supported by a shared commitment to fostering creativity and intellectual exploration. Public institutions, schools, and community centers embrace a model of education that values curiosity, critical thinking, and environmental awareness, preparing future generations to navigate a world where abundance is balanced with responsibility.

In this cultural renaissance, knowledge-sharing becomes a cornerstone of progress. Educational initiatives promote the principles of sustainability and ethical energy use, instilling an understanding of energy abundance as a shared gift rather than an individual possession. Communities come together to celebrate the beauty of balance and the power of restraint, embracing a collective philosophy that prioritizes the well-being of people and the planet over material consumption. The result is a society where each individual feels empowered to contribute to the common good, where knowledge is freely exchanged, and where progress is measured not by economic growth alone but by the health, happiness, and resilience of communities.

The legacy of an energy-abundant society is one that reaches far beyond technology and resource management. It represents a paradigm shift in how humanity perceives its place in the world and its responsibilities toward others and the environment. This legacy offers a vision of civilization at its best, where progress is driven by compassion, innovation, and respect for the interconnectedness of all life. Humanity, once bound by the limitations of finite resources, finds itself empowered to create a future that honors the dignity of individuals, the diversity of cultures, and the beauty of the natural world.

As the world transitions into this era of energy abundance, the journey itself becomes a source of collective strength. Each step forward reinforces a shared commitment to values that transcend individual gain, fostering a sense of unity and purpose that guides humanity toward a sustainable, equitable, and harmonious future. This journey is more than a technological revolution; it is a cultural, ethical, and philosophical evolution that redefines what it means to thrive as a global society. Energy abundance serves as a beacon, illuminating a path toward a world where human potential is unleashed not in isolation but in partnership with the planet, where every advancement contributes to a legacy of balance and stewardship.

In the face of this transformative vision, society stands united in its purpose, inspired by the possibilities of a world where energy scarcity is a relic of the past. With each breakthrough and each act of stewardship, humanity moves closer to a future that reflects the highest aspirations of civilization. The story of energy abundance is one of hope, resilience, and renewal, a testament to the power of collective action and shared responsibility. This legacy will shape the course of history, guiding future generations toward a world defined not by the limits of resources but by the boundless potential of ethical progress.

The journey toward energy abundance is one of both external transformation and internal growth. At its core, this transition redefines civilization's relationship with power, positioning energy not merely as a tool for consumption but as a force for creation, discovery, and restoration. The technological marvels that make energy abundance possible—superconductivity, zero-gravity energy generation, and wireless transmission—are only the physical manifestations of a deeper shift in consciousness. Humanity is called to recognize that, in harnessing boundless energy, it holds a gift and a responsibility, a dual mandate to elevate human potential while honoring the delicate balance of ecosystems and the planet's intrinsic beauty.

The availability of sustainable, limitless energy allows humanity to reach heights once limited by scarcity, bringing with it opportunities to reimagine every facet of society. Science, unburdened by energy constraints, becomes a field of infinite possibility. Researchers can explore frontiers in medicine, artificial intelligence, quantum computing, and space exploration with a vigor unrestrained by the practical limitations of finite resources. The ambition to understand and transform the world around us is fueled by an energy foundation that encourages inquiry, experimentation, and innovation on a scale previously unimaginable. In this era, the great challenges of human knowledge are met with curiosity and courage, as abundant energy enables scientists and visionaries to devote resources and intellect toward breakthroughs that benefit not only individuals but all of humanity.

Space exploration, once the province of superpowers and resource-rich nations, becomes a collective endeavor supported by the abundance of renewable energy. The power to venture beyond Earth's confines no longer depends on the enormous costs and environmental tolls associated with fossil fuels; it becomes a matter of curiosity, collaboration, and shared exploration. Humanity is no longer bound to its terrestrial limitations, able to pursue interplanetary exploration and settlement. Space habitats, powered sustainably, open pathways to interstellar knowledge, encouraging humanity to view itself as part of a larger cosmic family. This grand journey, empowered by energy abundance, shifts humanity's perspective from one of earthly dominion to universal participation, a profound recalibration that fosters both humility and reverence for the cosmos.

Back on Earth, the energy revolution fosters a cultural renaissance. With the removal of barriers imposed by scarcity, individuals are free to engage in pursuits that nurture creativity, innovation, and the arts. This abundance grants people the security to pursue vocations that may not yield immediate economic return but which contribute immeasurably to the richness of human experience. Artists, musicians, writers, and thinkers thrive in a society where energy abundance supports institutions and platforms that elevate the creative spirit. The arts become a celebrated facet of public life, inspiring individuals to look beyond their immediate needs and find beauty in expressions that transcend time, culture, and individual experience. This cultural flourishing fosters a sense of unity, reminding people of their shared humanity and the power of creativity to bridge divides.

Education, too, is transformed. The transition to energy abundance dissolves economic and logistical barriers that once limited access to learning. Schools, universities, and research centers operate without the constraints of energy scarcity, allowing them to function as hubs of innovation and growth. Students are encouraged to think expansively, developing skills and perspectives that contribute to a global society rooted in the principles of sustainability, ethics, and equity. Learning becomes a lifelong journey, accessible to all and available in every corner of the world, bridging gaps that once existed between urban and rural, affluent and underserved. With the support of abundant energy, virtual classrooms, digital libraries, and educational initiatives empower individuals to participate fully in the global community, regardless of geographical or economic limitations.

The economic impact of energy abundance reaches deeply into every sector, creating a foundation for sustainable industries and responsible growth. The manufacturing sector, once heavily reliant on finite resources, evolves into a model of closed-loop production where waste is minimized, materials are repurposed, and every component is designed for durability. This circular economy reduces environmental impact and fosters resilience, as industries no longer extract resources at unsustainable rates but instead create products that are part of a regenerative cycle. In an energy-abundant world, production is no longer a process of extraction and exhaustion but of renewal and restoration, transforming the very nature of industry into one that respects and preserves the natural environment.

Agriculture, one of humanity's oldest and most essential industries, also benefits profoundly from this shift. Sustainable energy supports innovations in vertical farming, aquaponics, and regenerative agriculture, enabling food production that nourishes people without depleting soils or water resources. Rural areas, once constrained by access to power, flourish as energy abundance enables advanced irrigation, climate-resilient crops, and local food production. This shift reduces the need for long-distance transportation of food, decreasing carbon emissions and supporting local economies. In this reimagined agricultural landscape, communities become self-sustaining, enjoying food security and independence while maintaining ecological balance.

Energy abundance also brings transformative change to global health systems. In a world with unlimited clean energy, healthcare facilities operate at peak efficiency, providing consistent, high-quality care to populations that once faced energy-related limitations. Advanced medical technologies, from imaging devices to telemedicine platforms, are accessible to all, bridging the healthcare gap between urban centers and remote regions. With energy no longer a constraint, medical research accelerates, driving discoveries in genomics, personalized medicine, and preventive care. These advancements extend life expectancy and improve quality of life across demographics, creating a world where health and wellness are within reach for everyone, regardless of location or socioeconomic status.

The benefits of energy abundance extend beyond human needs to encompass a renewed commitment to environmental stewardship. The elimination of fossil fuels as a primary energy source allows ecosystems to recover, enabling humanity to address the environmental damage accrued over centuries. Forests, which absorb and store carbon, become areas of preservation and expansion, reducing atmospheric CO_2 and supporting biodiversity. Polluted rivers and lakes are rehabilitated as clean energy powers water treatment systems, returning vital ecosystems to health. Urban areas, once plagued by air pollution and smog, are revitalized as the sources of emissions are eliminated, creating cities where clean air and green spaces become the standard rather than the exception. The natural world, no longer seen merely as a resource to exploit, is recognized as an intricate system that sustains all life, fostering a newfound respect for ecological integrity.

In this energy-abundant society, humanity is invited to reflect on its ethical obligations. While technology provides the framework for limitless energy, it is human choice that determines its application. This newfound power calls for a guiding philosophy that prioritizes equity, compassion, and responsibility. Energy abundance must not become a tool of control or oppression; rather, it should serve as a foundation for a more just and equitable society. To achieve this, governments, institutions, and individuals must commit to principles that uphold human rights, promote access to resources, and ensure that the benefits of energy are shared inclusively.

As global governance adapts to this new reality, there is an emphasis on transparency and cooperation. Policies and regulatory frameworks are established to prevent monopolization and safeguard energy as a public good. International organizations take on the role of ensuring fair access and ethical practices, providing a structure that supports equitable distribution and responsible management. By upholding these standards, humanity creates a world where energy abundance supports freedom, opportunity, and equality, empowering individuals to lead lives defined by dignity and purpose. This vision of responsible governance fosters a society that places collective well-being at its core, shaping laws and institutions that reflect humanity's highest values.

Culturally, the impact of energy abundance is transformative. Freed from the constraints of scarcity, communities develop a collective mindset centered on cooperation and sustainability. Society embraces a cultural ethos that values simplicity and respect for the environment, recognizing that progress need not come at the expense of natural balance. A culture of mindfulness emerges, where people value restraint and deliberate use of resources, guided by the understanding that true prosperity lies in harmony rather than excess. This cultural renaissance encourages people to view themselves as part of an interconnected whole, fostering an awareness that the well-being of one community depends on the health of the planet as a whole.

The transition to energy abundance enables a reimagining of infrastructure, where cities and communities are built not just to function but to flourish within their ecosystems. Urban planning incorporates renewable energy into every aspect of design, creating cities that are as much a part of the natural landscape as the forests and rivers that surround them. Transportation networks operate on clean energy, reducing pollution and enhancing connectivity. Public spaces are designed to foster community engagement and environmental education, creating areas where people can come together to learn, create, and celebrate. In these cities, architecture and technology blend seamlessly with nature, embodying the principles of resilience, beauty, and sustainability.

Energy abundance also changes how humanity perceives time and legacy. With sustainable energy supporting the essentials of life, society can shift its focus from short-term gain to long-term impact. Projects that require decades or even centuries to complete are pursued with confidence, unbound by the constraints that once forced rapid, unsustainable growth. This new relationship with time encourages an approach to progress that is patient and thoughtful, emphasizing the importance of creating a world that will thrive not just for present generations but for those yet to come. Humanity finds itself inspired to leave a legacy of balance and beauty, choosing to build not only for itself but for the enduring health of the planet.

In this new era, the question of what it means to be human is revisited. Energy abundance does not simply enhance life; it changes its essence, allowing individuals to pursue lives defined by purpose, creativity, and community. Freed from the pressures of survival, people are empowered to cultivate their unique talents and explore their passions. The arts, sciences, and social institutions flourish, supported by an infrastructure that values human potential as a renewable resource in its own right. The collective pursuit of knowledge, beauty, and innovation becomes the measure of a successful society, where the fruits of abundance are shared to inspire and elevate everyone.

Ultimately, the story of energy abundance is one of transformation—a testament to humanity's capacity to reimagine its relationship with the planet and with each other. It is a journey toward a world where progress aligns with ethical and environmental integrity, where the boundless potential of renewable power serves not as an end but as a means to foster unity, resilience, and hope. In this vision, energy abundance illuminates a path to a future where humanity's greatest achievements are not measured by consumption but by its contributions to a world defined by peace, balance, and shared prosperity. This legacy, born of energy abundance, is one that celebrates not only the power of technology but the enduring strength of compassion, integrity, and wisdom.

As humanity completes its transition into an era of energy abundance, the full magnitude of this transformation unfolds, redefining not only the external world but also the internal dimensions of human consciousness, community, and purpose. In a society unburdened by the limitations of scarcity, civilization flourishes in ways that transcend traditional metrics of progress, ushering in a new ethos centered on stewardship, harmony, and shared prosperity.

The profound impact of energy abundance reshapes society from the ground up, fostering a world where the struggle for resources no longer dictates relationships, and where energy is a common good accessible to all. This shared access dissolves historical divides, fostering an interconnected global community where cooperation and mutual support become the guiding principles of progress. Nations, once divided by competition for finite resources, now find common ground in shared goals of sustainability, environmental renewal, and equitable growth. Humanity, in embracing energy abundance, transcends the zero-sum games of the past and embarks on a collective journey toward a future where success is measured not by consumption but by the quality and resilience of the ecosystems, communities, and cultures that sustain life on Earth.

This transformation demands a philosophical evolution, a redefinition of what it means to thrive as a civilization. Energy abundance offers a unique opportunity for humanity to move beyond the cycles of extraction, growth, and depletion that have historically defined progress. In this new paradigm, growth is regenerative, aligning with the rhythms of nature and prioritizing longevity over short-term gain. Each innovation, each technological advancement, each societal achievement is designed with an awareness of its impact on future generations. This intergenerational perspective fosters a culture that values foresight and responsibility, where the pursuit of knowledge, creativity, and innovation is tempered by an understanding of humanity's role as custodians of the planet.

The ethical foundations of energy abundance extend beyond equitable access; they encompass a commitment to ecological harmony and respect for the intricate systems that sustain life. Freed from the constraints of fossil fuels, society no longer views the natural world merely as a resource to be harnessed but as a partner in a shared journey. This shift transforms humanity's relationship with the environment, fostering a profound respect for biodiversity, natural cycles, and the ecological webs that connect all species. The era of energy abundance catalyzes an ecological renaissance, where restoration and conservation efforts flourish, supported by sustainable power that enables large-scale environmental projects without the burden of pollution or resource depletion. Forests are restored, wetlands are protected, and endangered species find refuge as humanity turns its efforts toward healing the planet.

The benefits of this new relationship with the environment are not merely ecological; they are also spiritual and cultural. Humanity, freed from the demands of survival and scarcity, finds itself with the opportunity to reconnect with the natural world in ways that foster a deep sense of belonging, humility, and wonder. As societies embrace sustainable practices, they cultivate a reverence for the interconnectedness of life, nurturing a worldview that values harmony over dominion. Public spaces, urban parks, and natural reserves become sanctuaries of reflection, where individuals and communities can experience the beauty and tranquility of nature. This cultural shift fosters a sense of shared responsibility, a recognition that the well-being of humanity is intrinsically linked to the health of the planet.

Education, as the cornerstone of societal progress, evolves in this energy-abundant world to reflect these new values. Schools, universities, and learning centers embrace curricula that emphasize environmental literacy, ethics, and global citizenship, preparing future generations to navigate a world where responsibility, equity, and sustainability are the guiding principles. Students are encouraged to think critically, to question, to innovate, and to view knowledge not as a means of personal advancement but as a tool for collective improvement. This reimagined educational system produces individuals who are not only equipped to contribute to society but who understand their roles as stewards of a shared world. The legacy of energy abundance is thus passed on to future generations, ingrained in the minds and hearts of those who will inherit and continue to shape this world.

The arts, too, experience a renaissance in an energy-abundant society. With the freedom to create without the limitations of scarcity, artists, musicians, writers, and visionaries find themselves supported by communities that value their contributions to cultural richness. Art becomes a means of exploring humanity's relationship with the planet, expressing the complex emotions and insights that arise from this unprecedented era. Festivals, museums, and public exhibitions flourish, celebrating creativity as a shared endeavor that connects individuals across boundaries of language, geography, and culture. Art and creativity become central to public life, inspiring a sense of unity and reinforcing the shared values that guide society. This cultural renaissance serves as a reminder of humanity's capacity for beauty, resilience, and empathy, qualities that become ever more precious in a world that values harmony and balance.

In the realm of technology, the impact of energy abundance is equally transformative. Freed from the energy-intensive demands of traditional power sources, technology becomes a force for regeneration rather than consumption. Innovations in clean manufacturing, waste reduction, and sustainable production redefine industries, enabling economies to thrive without the environmental costs that once accompanied industrial growth. The fields of artificial intelligence, robotics, and biotechnology expand with renewed ethical frameworks, prioritizing applications that support health, sustainability, and social well-being. Technology is no longer seen as an instrument of extraction but as a tool for nurturing life, fostering resilience, and enhancing the quality of human experience. In this era, humanity's relationship with technology is marked by a spirit of humility and stewardship, recognizing that the true power of innovation lies not in domination but in the ability to enhance life for all.

Governance in an energy-abundant world is redefined by principles of transparency, equity, and collaboration. With the fundamental needs of society met, governments can prioritize policies that support well-being, environmental health, and cultural vitality. Public institutions are tasked with protecting the rights of all citizens, ensuring that energy abundance serves as a universal asset rather than a privilege. International cooperation becomes the foundation of global stability, as nations recognize that the shared benefits of energy abundance require shared responsibility. Diplomatic relations are characterized by a commitment to sustainability, peace, and mutual respect, fostering a world where collaboration replaces conflict and where resources are managed for the benefit of humanity as a whole.

The ethical principles guiding governance in this era encourage a rethinking of what it means to lead. Leaders are no longer defined by their capacity to amass power but by their dedication to service, their commitment to ethical progress, and their vision for a world that thrives through unity and sustainability. This transformation in leadership fosters a generation of policymakers, scientists, and community advocates who are deeply aware of their responsibility to future generations. They work not for personal gain but for the lasting well-being of society, creating a legacy of integrity and respect that stands as a testament to the ideals of this new age.

At the heart of this transformation lies a shift in the collective consciousness, a realization that energy abundance is not an end in itself but a means to a greater purpose. Humanity, with access to boundless power, is invited to reflect on the deeper questions of existence, purpose, and legacy. The energy to sustain life, foster innovation, and empower communities is a gift that calls for humility, gratitude, and a profound respect for the interconnectedness of all life. This awareness infuses society with a sense of purpose that transcends individual ambition, inspiring people to contribute to a future defined by balance, compassion, and stewardship.

The legacy of energy abundance, then, is not simply one of technological achievement; it is a legacy of enlightenment, a turning point in human history where civilization chooses to build a world that reflects the highest ideals of harmony, equity, and resilience. It is a legacy that future generations will inherit with pride, one that reminds humanity of its capacity to evolve not only technologically but ethically, culturally, and spiritually. The journey to energy abundance is a journey toward a more complete understanding of what it means to live in harmony with the planet and with each other, a journey that shapes the very essence of civilization.

As this journey unfolds, humanity stands united in its purpose, guided by the vision of a world where energy fuels not only machines but the collective aspiration to live in peace, justice, and unity. This era of energy abundance does not signify the end of progress but the beginning of a new chapter, one where the potential for human achievement is matched by a commitment to ethical stewardship. In embracing energy abundance, humanity takes its place as a responsible and compassionate force within the natural world, building a legacy that honors the beauty, diversity, and interconnectedness of all life.

In the final analysis, energy abundance is more than a solution to scarcity; it is a profound shift that touches every aspect of life, inviting humanity to reimagine what it means to flourish. This new age of shared prosperity, fueled by sustainable power, fosters a world where individuals are free to pursue knowledge, beauty, and purpose, where communities are resilient, and where the planet is cherished as a home for all. The story of energy abundance is a story of hope, a testament to humanity's capacity for growth, empathy, and foresight—a legacy that will inspire generations to come, as they too inherit the dream of a world where energy is not a barrier but a bridge to a more harmonious, compassionate, and enlightened future.

Chapter 1: The Path to Energy Abundance

At the dawn of humanity's understanding, energy was an ungraspable mystery, a force harnessed through fire, muscle, and wind. In these early days, energy's potential was little more than a concept embedded in the heat of the sun, the crash of waves, and the sporadic flicker of flames. But as civilizations rose and curiosity grew, so too did the awareness that this invisible power, coursing through nature and fueling human endeavor, was a cornerstone of progress, shaping every significant leap forward. From the fires that warmed ancient settlements to the windmills that powered small villages, energy transformed from an intangible entity to a purposeful, albeit rudimentary, tool.

As the centuries turned, humanity's mastery over energy surged. The industrial era, with its steam engines and coal furnaces, marked a point of no return. This epoch ignited a transformation of the human experience, accelerating production, fueling mechanized agriculture, and fundamentally reshaping society. Where once energy had been a gentle, occasionally capricious force, it now roared through factories and railways, asserting itself as an indispensable engine of progress. And yet, the demands of this newly industrialized world came at a cost. The fossil fuels that fed the fires of industry were not infinite, and the air that grew thick with soot carried warnings that went unheeded for generations. But it was this very period, as humanity plunged headlong into an age of mechanization, that laid the foundation for the next stage: the realization that energy itself needed reimagining if civilization were to continue its march forward.

The advent of electricity represented a paradigm shift, illuminating entire cities and changing the fabric of daily life. Energy was no longer confined to the physical exertion of humans or animals, nor the limitations of steam and coal. It was now invisible, powerful, and seemingly limitless, flowing through wires to reach every corner of urban existence. This period was transformative, democratizing access to power and spreading the benefits of industrialization beyond the confines of factories to homes, schools, and hospitals. Electricity promised to elevate the human condition, illuminating homes, driving motors, and connecting people across vast distances. Yet this promise carried with it a silent compromise: reliance on fossil fuels intensified, and the effects of extraction and combustion began to mount, casting shadows on the bright lights of progress.

As the twentieth century dawned, the reach of energy expanded beyond mere utility. Scientists and visionaries recognized its potential to transform not only industry but the very essence of human interaction, comfort, and ambition. With each new discovery, from the wonders of nuclear power to the delicate balances of chemical batteries, the boundaries of what could be achieved shifted once again. Nuclear energy, though fraught with risks, hinted at the notion of boundless power, of an energy source that could theoretically supply the world's needs many times over. Meanwhile, batteries enabled mobility, the freedom to carry power itself in one's hands, liberating technology from fixed infrastructure. Energy, now more versatile and dynamic than ever, moved into a realm that was as much about imagination and potential as it was about practical applications.

Yet, as technology advanced and humanity's thirst for energy intensified, a critical realization set in. Energy, despite its myriad forms and applications, remained tethered to limitations. Fossil fuels, once seemingly endless, began to dwindle, revealing their ecological toll in polluted rivers, poisoned air, and warming oceans. The promise of nuclear power was tempered by the haunting specter of accidents and radioactive waste, a chilling reminder of the fine line between advancement and catastrophe. Even renewable sources, though celebrated, presented challenges of their own. The sun does not always shine; the wind does not always blow. Humanity was confronted with the harsh truth that energy, in its current forms, was bounded by nature's own rhythms and reserves.

This period of reckoning ignited a new wave of thought, a shift toward rethinking energy not merely as a means to an end but as a complex, interwoven system that must balance both human needs and environmental imperatives. Scientists, policymakers, and citizens alike began to recognize that the search for energy solutions could no longer be a short-sighted endeavor. It was a journey that demanded foresight, innovation, and a willingness to transcend the immediate in pursuit of a sustainable future. The promise of abundance—a world where energy flows freely, renewably, and equitably—became both a necessity and an aspiration.

And so began the pursuit of renewable energy, a movement fueled not only by technological advances but by a growing awareness of humanity's impact on the planet. Solar, wind, geothermal, and hydropower emerged as the harbingers of a new era, a future where energy could be derived from natural forces that had powered the Earth long before human civilization. But despite their promise, these sources remained constrained by the same forces that offered them: the sun's energy, however powerful, is intermittent; the winds, however forceful, are inconsistent. This reality introduced a paradox that modern society had to confront: the sources of energy most compatible with nature were also those most bound by it.

It was in this paradox that the vision of energy abundance found its roots. What if humanity could overcome these limitations? What if the dream of boundless, reliable, and environmentally harmonious energy could be realized? Such questions moved beyond theoretical debate, inspiring new fields of research and shifting the focus of energy policy worldwide. The vision of a world where energy was not rationed but compounded, where power did not compete with the environment but reinforced it, became a unifying ideal—a vision of not just survival, but thriving.

In seeking to fulfill this vision, humanity turned to the skies, to the vast, unyielding potential of outer space. Space, a realm untouched by terrestrial constraints, offered an arena where gravity and atmospheric drag were no longer obstacles. The frictionless environment of space opened new possibilities for energy generation, enabling technologies that could operate with efficiency impossible on Earth. Scientists proposed orbital solar arrays that could capture the sun's rays without interruption, producing a steady stream of power and beaming it down to Earth. This concept, though ambitious, promised an escape from the cycle of scarcity, presenting a future where energy could flow uninterrupted by the limitations of weather or terrestrial geography.

Yet this vision extended beyond the mere capture of solar energy. It touched upon the idea of energy compounding, of creating systems that not only generated power but also reinvested it, growing in capacity over time. Just as financial wealth could compound through reinvestment, so too could energy multiply through cycles of self-sustenance. Superconductors, which allow electricity to flow without resistance, hinted at the possibility of achieving this compounding effect. In the cold, vacuumed environment of space, these superconductors could operate at maximum efficiency, transforming theoretical possibilities into practical realities.

The prospect of superconducting power systems operating in space brought humanity closer to the dream of perpetual energy. These systems could, in theory, operate indefinitely, generating a surplus that could be transmitted back to Earth without the constant need for input. What emerged was the vision of a world liberated from the constraints of fossil fuels and finite resources, a world where energy not only sustained life but enabled it to flourish without the shadows of depletion and degradation. This was the promise of energy abundance, not as an abstract ideal but as a tangible pathway toward a more balanced, sustainable world.

Yet, as the vision of energy abundance grew, so too did the questions it raised. How would such a world be governed? Could humanity handle the responsibility of boundless energy without falling into the same patterns of exploitation and inequality that had marred previous eras? The shift toward energy abundance was not just a scientific or technological endeavor; it was an ethical and philosophical transformation. This new era called for a rethinking of values, an awareness that energy abundance could either uplift humanity to new heights or deepen existing divides.

The journey toward energy abundance thus became a dual pursuit: a scientific endeavor rooted in engineering and physics, and a moral undertaking that required humanity to confront its own values. Would energy abundance become a tool for unity or a source of division? Would it foster a sense of shared stewardship or become a force of control wielded by the few? These questions underscored the need for a holistic approach, one that recognized energy not as a commodity to be owned but as a shared resource, a foundation upon which a fair and flourishing society could be built.

As humanity's understanding of energy evolved, so too did the tools and methods by which it was harnessed. Early dreams of abundant energy hinged on visions of nuclear fusion and vast hydroelectric dams, but these aspirations, while groundbreaking, ultimately revealed limitations both practical and moral. Nuclear power, despite its vast potential, became a sobering reminder of the delicate balance between progress and risk, while large-scale dams, though impressive, disrupted ecosystems and displaced communities. These challenges underscored a need for a new paradigm—one that did not merely exploit nature's resources but harmonized with them, creating a synergy between technological advancement and ecological responsibility.

It was this search for balance that brought scientists and thinkers to explore the untapped potential of the natural world, not merely as a source of raw materials but as a model of efficiency and regeneration. Nature, with its ecosystems of energy flows and closed loops, offered profound insights into sustainable power. Plants, capturing sunlight and converting it to energy, operated on principles that were both efficient and renewable. Ecosystems thrived through cycles that recycled every molecule, every bit of energy, so that nothing was truly wasted. These natural processes inspired a vision where energy systems could mirror nature's own cycles, creating self-sustaining systems that enriched rather than depleted.

The concept of energy compounding was born from this vision, a notion that energy could be made to regenerate, to cycle through systems in a way that magnified its output. Just as the financial principle of compounding turns small returns into exponential growth, energy compounding promised a world where each surplus of power reinvested itself into the system. This would create a cascade of energy generation, where output continuously amplified over time, enabling a steady flow of power that could match, and even exceed, humanity's demands. This vision of energy compounding was no longer a distant dream; it was an emerging reality, fueled by breakthroughs in superconductivity, zero-gravity environments, and advanced transmission methods.

Space, with its frictionless environment and boundless solar exposure, emerged as the ideal frontier for this new energy model. In space, solar arrays could capture the sun's energy without interruption, free from the cyclical patterns of night and day or the interference of weather. Scientists envisioned massive solar stations in orbit, panels stretching across miles, soaking up sunlight and converting it into a constant stream of power. This energy, generated continuously, could be transmitted back to Earth using advanced wireless technologies, ensuring that power reached even the most remote and underserved regions.

The potential of superconductivity to complement this model added a layer of transformative promise. Superconductors, materials that allow electricity to flow without resistance, had long captivated the scientific community. On Earth, where temperatures rarely dip to the levels required for superconductivity, the applications of these materials remained limited. But space, with its cold, near-zero environments, offered a perfect setting for superconductors to operate at peak efficiency. By coupling solar stations with superconducting materials, scientists envisioned a system where energy could flow continuously, without loss, creating a circuit that was both unbreakable and self-sustaining.

In this envisioned future, the energy generated in space would flow seamlessly back to Earth, transmitted through wireless technology inspired by the early work of Nikola Tesla, who had envisioned a world connected by wireless power long before the technology existed to make it feasible. Today's advancements in resonant inductive coupling and microwave transmission have brought this vision closer to reality, with scientists exploring ways to beam energy across vast distances with minimal loss. With energy flowing freely from space-based solar arrays to Earth, the very structure of terrestrial power grids could be reimagined, creating networks that are both resilient and inclusive, reaching every corner of the globe.

But the vision of energy abundance extended beyond technical feasibility. It posed a profound question about humanity's role as both creator and steward, challenging societies to consider what such power would mean for the future of civilization. In a world where energy scarcity no longer dictated the terms of progress, what new possibilities could emerge? Would humanity use this newfound abundance to uplift all people, fostering a world where access to energy was a universal right? Or would the specter of inequality and control continue to haunt this new frontier, as energy monopolies and geopolitical rivalries adapted to maintain dominance?

These questions became central to the movement toward energy abundance, underscoring the need for an ethical framework as robust as the technologies that supported it. Without ethical stewardship, energy abundance could become a source of discord rather than harmony, a tool of division rather than empowerment. It became clear that this vision required more than scientific innovation; it demanded a commitment to fairness, to the protection of ecosystems, and to the equitable distribution of resources. This holistic approach was essential to ensure that energy abundance would fulfill its promise not just as a technological marvel but as a catalyst for social and environmental progress.

In contemplating a world fueled by abundant energy, the promise of equity rose to the forefront. Energy scarcity had long been a dividing line, separating those with access from those without, often reinforcing cycles of poverty and limiting opportunities for growth and development. The dream of energy abundance was not simply to eliminate scarcity but to do so in a way that democratized access, reaching communities that had long been overlooked. By removing energy as a barrier, societies could address deeper inequalities, giving people the freedom to innovate, to learn, to connect, and to grow. The vision of energy abundance was, at its heart, a vision of social transformation.

This transformation would reshape every industry and sector, creating a foundation for sustainable, inclusive growth. Manufacturing, often limited by energy costs and supply-chain constraints, would find new efficiencies in renewable power, reducing environmental impacts while increasing productivity. Agricultural practices could be reimagined to incorporate energy-intensive innovations like vertical farming and hydroponics, producing food with less land and water. The healthcare industry, too, would flourish, with hospitals and clinics in remote areas empowered by reliable, renewable energy that enabled advanced equipment and telemedicine, bridging the gap between urban and rural health services. In an energy-abundant world, the limitations that once stifled development in these sectors would no longer hold sway.

Education, often constrained by geography and resources, would benefit profoundly from this transformation. With universal access to energy, schools, universities, and research institutions could operate with continuity and creativity, connecting students and teachers across vast distances. This would create a global community of learners, where knowledge and resources could be shared freely, leveling the playing field and fostering an era of intellectual exchange and discovery. Education would no longer be limited by physical boundaries or economic disparities; it would be a right accessible to all, supported by an infrastructure that recognized the intrinsic value of learning.

As societies reaped the benefits of energy abundance, a cultural shift emerged, one that viewed energy not as a commodity to be consumed but as a resource to be respected and sustained. This shift was underpinned by an ethic of conservation, a recognition that even in abundance, resources should be used wisely and mindfully. Public campaigns, educational programs, and community initiatives fostered a culture of awareness, where people were encouraged to consider the environmental impacts of their energy use. Conservation became a shared value, embedded in the fabric of daily life, a reminder that abundance did not justify waste. In this way, energy abundance did not foster excess; it fostered responsibility.

This ethic of conservation extended beyond energy to encompass the entirety of humanity's relationship with the planet. With energy abundance, societies had the power to not only halt environmental degradation but to reverse it. Initiatives in reforestation, habitat restoration, and species protection became central to public policy, powered by renewable energy that enabled large-scale projects without harming the environment. Water purification systems transformed polluted rivers into lifelines for communities and wildlife. Renewable-powered desalination plants made fresh water accessible to arid regions, supporting agriculture and reducing stress on freshwater ecosystems. These efforts marked a shift from a world focused on mitigating damage to one actively engaged in healing the planet.

The environmental transformation fueled by energy abundance extended into urban planning and infrastructure. Cities, long defined by their resource dependencies, began to evolve in ways that minimized environmental impact while maximizing quality of life. Buildings integrated solar panels, green roofs, and sustainable materials, reducing energy consumption and lowering carbon footprints. Public transportation systems, powered by renewable energy, provided efficient, low-impact alternatives to traditional vehicles, reducing pollution and easing traffic congestion. Parks and green spaces became central features of urban design, creating oases of biodiversity in the heart of metropolitan areas. In these cities, nature and technology coexisted in harmony, each complementing the other in a balanced, integrated system.

Through these transformations, energy abundance became a cornerstone of cultural and societal renewal. Freed from the relentless pursuit of resources, communities could invest in the arts, in public spaces, and in social services that fostered well-being and connection. Festivals, museums, and theaters flourished, bringing people together in celebration of creativity and shared experience. The arts, once constrained by financial and logistical limitations, thrived in an environment where innovation was encouraged and supported. This cultural renaissance enriched society, creating a shared language of beauty and expression that united people across diverse backgrounds and perspectives.

The vision of energy abundance reshapes not only the physical world but also the underlying values and principles that guide humanity's relationship with power, progress, and the environment. With the pursuit of a society no longer bound by the cycles of scarcity, every element of life is touched by the transformative potential of sustainable energy. Freed from traditional constraints, civilization gains the ability to address longstanding challenges, not by temporary solutions but through a holistic restructuring that elevates the human experience while preserving the planet.

In this new paradigm, energy abundance brings about a shift in how societies define prosperity. No longer rooted in material accumulation or extractive practices, prosperity becomes synonymous with balance, creativity, and resilience. The idea of growth, often linked to consumption, is reimagined in terms of sustainable development and long-term impact. This shift invites every sector, every institution, and every individual to consider the ethical dimensions of their actions, recognizing that true progress is measured by its contributions to collective well-being rather than individual gain. This redefinition of prosperity encourages societies to embrace values of stewardship, equity, and compassion, creating a foundation for an age defined not by scarcity but by ethical advancement.

The ethical dimensions of energy abundance also influence economic structures, reshaping the foundations of global commerce and industry. In a world no longer driven by competition over limited resources, businesses are encouraged to adopt models that prioritize sustainability, fair labor practices, and environmental responsibility. The economy becomes an ecosystem where companies thrive by contributing to the common good rather than merely extracting value. Circular economies, in which waste is minimized and materials are repurposed, flourish under this model, reducing environmental impact while fostering resilience. Industries that once relied on resource exploitation now focus on regeneration and longevity, producing goods that are designed to be reused, repaired, and recycled. This approach not only reduces waste but aligns economic activity with the natural cycles of the Earth, creating a harmonious balance between production and preservation.

For individuals, energy abundance opens doors to new opportunities for self-actualization and fulfillment. With energy no longer a limiting factor, people are empowered to pursue vocations and interests that contribute to personal growth and community development. Freed from the pressures of scarcity, individuals have the freedom to explore creative, intellectual, and philanthropic pursuits, nurturing a society where human potential is fully realized. The abundance of energy transforms work from a means of survival to a source of purpose, where people engage in activities that resonate with their values and passions. This new approach to work fosters a sense of fulfillment and connection, enriching both individual lives and the broader community.

Education, the cornerstone of societal advancement, benefits profoundly from this shift in priorities. In an energy-abundant world, learning becomes a lifelong endeavor, accessible to all and supported by a global infrastructure that values knowledge as a public good. Schools and universities, unburdened by energy costs, become centers of exploration and innovation, fostering a culture that celebrates curiosity and critical thinking. Advanced digital platforms and virtual classrooms connect students and teachers across continents, creating a global network of shared knowledge and collaborative learning. This educational transformation empowers individuals to reach their full potential, preparing them to contribute to a society that values not only technical skills but also ethical awareness and environmental responsibility.

This new era of education also emphasizes the importance of interdisciplinary learning, encouraging students to view the world through multiple lenses—scientific, artistic, philosophical, and ecological. By integrating these perspectives, individuals develop a holistic understanding of their place within the larger systems of society and the environment. Education becomes a means of cultivating empathy, resilience, and a sense of shared purpose, equipping future generations with the tools to navigate an interconnected world where their choices have far-reaching implications. In this way, the energy-abundant society fosters not only intellectual growth but also moral and emotional maturity, creating a legacy of informed, compassionate citizens who are prepared to lead with integrity and foresight.

The healthcare sector, too, undergoes a profound transformation in an energy-abundant society. Access to reliable, clean energy enables healthcare facilities in even the most remote regions to operate advanced medical equipment, maintain optimal environmental controls, and provide continuous care. Telemedicine becomes a standard service, connecting patients with specialists regardless of geographic barriers, while mobile clinics bring healthcare to underserved populations. This accessibility democratizes healthcare, ensuring that every individual has the opportunity to receive quality care and live a healthy life. Medical research also accelerates, fueled by the reliable energy needed to power cutting-edge laboratories and data centers. Innovations in genomics, regenerative medicine, and disease prevention benefit from this infrastructure, extending life expectancy and enhancing the quality of life across demographics.

The environmental implications of energy abundance extend beyond pollution reduction to encompass a full-scale commitment to ecological restoration. Freed from the necessity of fossil fuel extraction, societies can redirect resources toward rewilding projects, habitat conservation, and climate resilience efforts. Forests, once cleared for fuel or farmland, are allowed to regenerate, creating carbon sinks that help stabilize the climate. Oceans, free from the pollutants of oil spills and plastic waste, begin to heal, restoring marine ecosystems that are vital to global biodiversity. Wetlands, prairies, and river systems are rehabilitated, allowing natural landscapes to regain their integrity and support diverse forms of life. This era of ecological renewal redefines humanity's role on Earth, transforming people from exploiters to guardians of the planet.

In urban areas, the principles of energy abundance inspire new approaches to city planning and architecture. Cities are designed to be self-sustaining, with integrated renewable energy systems, green spaces, and efficient public transportation networks. Vertical farms and rooftop gardens bring food production into the urban environment, reducing reliance on long-distance supply chains and lowering the carbon footprint of food consumption. Smart buildings, equipped with energy-efficient technologies, optimize energy use and create healthier indoor environments. Public spaces are redesigned to foster community engagement and environmental awareness, providing citizens with places to connect, learn, and collaborate. These energy-abundant cities are not only functional but beautiful, embodying a vision of urban life that harmonizes with nature and promotes well-being.

The transformative potential of energy abundance also reshapes transportation, one of the most significant contributors to carbon emissions. Renewable energy powers an extensive network of electric public transportation, reducing reliance on personal vehicles and lowering pollution levels. High-speed trains connect distant regions, enabling efficient, sustainable travel that bridges urban and rural areas. Personal electric vehicles, charged by abundant clean energy, become affordable and accessible to all, reducing the need for fossil fuels while empowering individuals with mobility. In this reimagined transportation landscape, movement is not a strain on the environment but a seamless, sustainable aspect of daily life.

As humanity embraces this new model of living, cultural values evolve to reflect a commitment to balance and interconnectedness. Energy abundance encourages a mindset that views resources as shared assets, fostering a sense of solidarity and responsibility toward one another and the planet. Communities come together to celebrate sustainable practices, from energy conservation to local food production, creating a culture that values mindfulness and intentionality. This shift in values permeates every aspect of society, from personal choices to public policy, reinforcing the idea that progress is measured not by individual gain but by collective harmony and resilience.

In this cultural transformation, art and storytelling play a central role, capturing the aspirations, challenges, and triumphs of a world in transition. Artists, writers, and filmmakers explore themes of environmental stewardship, social justice, and human potential, creating works that inspire reflection and action. Public art projects, festivals, and community events celebrate the beauty of a balanced, sustainable world, fostering a sense of pride and shared identity. Art becomes a bridge between generations, passing down the values of energy abundance and ecological harmony to future leaders and thinkers. In this way, culture becomes both a reflection of and a guide for the values of an energy-abundant society, shaping collective consciousness and inspiring continued growth.

At the heart of this transformation lies a reimagined relationship between humanity and technology. In an energy-abundant world, technology serves not as a means of exploitation but as a tool for preservation, restoration, and enhancement. Innovations in artificial intelligence, biotechnology, and environmental engineering are guided by ethical frameworks that prioritize sustainability, equity, and well-being. Technology becomes a partner in humanity's journey toward resilience, supporting efforts to conserve resources, protect ecosystems, and improve quality of life. This ethical approach to technology ensures that advancements align with the values of a society that recognizes its responsibility to the planet and to each other.

The governance of energy abundance, too, requires a rethinking of traditional structures. In a world where energy is both abundant and renewable, governments are tasked with ensuring fair access, ethical management, and international cooperation. Public policy prioritizes the needs of all citizens, creating frameworks that protect the environment, uphold human rights, and prevent monopolization. International alliances foster shared research, technological exchange, and collaborative problem-solving, building a foundation of global solidarity. By governing energy abundance with integrity and transparency, leaders create a world where power is not a source of conflict but a pathway to peace and unity.

This era of energy abundance is a testament to humanity's capacity for growth and transformation. It is a story of resilience, a journey from scarcity to sustainability that redefines what it means to progress as a civilization. In embracing renewable energy and sustainable practices, humanity lays the groundwork for a legacy of harmony, a world where the pursuit of knowledge, creativity, and justice is supported by the energy to bring these ideals to life. Energy abundance becomes more than a technological achievement; it is a philosophical awakening, a shift toward a worldview that values balance, respect, and shared purpose.

In this vision, the story of energy abundance is also a story of human potential. With the barriers of scarcity removed, individuals are empowered to pursue their highest aspirations, from scientific exploration to artistic expression, from personal growth to community service. This empowerment is not limited to the privileged few but extends to all, creating a society where everyone has the opportunity to contribute to and benefit from the collective journey. This new era reaffirms the value of each person's unique gifts and perspectives, fostering a society that celebrates diversity, inclusivity, and mutual respect.

As humanity embraces the possibilities of energy abundance, it stands at the threshold of a future defined not by limitations but by boundless potential. This future is built on the foundation of sustainability, guided by the principles of equity, and inspired by a commitment to leave the world better than it was found. In choosing to pursue energy abundance, humanity makes a promise to future generations—a promise to protect the planet, to uplift each other, and to honor the interconnectedness of all life. This promise is the legacy of energy abundance, a legacy that will endure as a testament to humanity's vision, integrity, and compassion.

The vision of an energy-abundant world invites humanity to dream of a future unbound by the constraints that have defined the past. This future is not one of passive consumption but active participation, where each individual, community, and nation contributes to a global tapestry of interconnected resources and shared knowledge. Energy abundance redefines the relationship between people and the planet, fostering a world where sustainable practices and ethical considerations shape every decision. This world encourages individuals and societies alike to rise to their highest potential, propelled by the understanding that power, when used responsibly, is a force for peace, prosperity, and harmony.

In this reimagined world, the principles of sustainability and resilience permeate every facet of life. Energy abundance is not viewed as an invitation to unchecked growth but as an opportunity to adopt a balanced approach that respects natural limits and cultivates regenerative practices. This era marks a departure from the cycles of extraction and depletion that have characterized human progress for centuries. Instead, humanity embraces a model of development that nurtures the Earth's ecosystems, recognizing that true abundance is achieved only when nature thrives alongside human civilization.

The benefits of this regenerative approach are profound, transforming agriculture, urban infrastructure, and industry alike. Agriculture, for instance, evolves beyond conventional farming methods, embracing innovations that reduce environmental impact and increase resilience. With an ample supply of clean energy, vertical farms flourish, bringing fresh produce to urban centers without the need for long-distance transport or extensive land use. Hydroponic and aquaponic systems, powered by renewable energy, offer high-yield solutions that conserve water, enhance food security, and support sustainable practices. Farmers incorporate precision agriculture, using data-driven insights to optimize crop growth and minimize waste. This new model of agriculture aligns with the rhythms of the Earth, producing food in harmony with the environment and ensuring that future generations inherit a land that is as fertile and abundant as it is today.

Urban life, too, is transformed by the principles of energy abundance. Cities no longer sprawl endlessly, consuming vast tracts of land and resources; instead, they evolve into compact, efficient ecosystems that blend seamlessly with their surroundings. Buildings are designed to be energy-positive, generating more power than they consume, while public spaces are designed to foster community, biodiversity, and environmental awareness. Parks, gardens, and green rooftops bring nature into the heart of the city, providing habitats for wildlife and spaces for people to connect with the natural world. Public transportation, powered entirely by renewable energy, connects every part of the city, reducing pollution and making it easier for residents to live sustainably. In these energy-abundant cities, quality of life is no longer tied to economic wealth alone but to the well-being of people and the health of the environment.

The ripple effects of these changes extend far beyond city limits, fostering a global shift toward responsible resource management and environmental stewardship. Industries, once focused on profit-driven extraction, now prioritize practices that support regeneration and ecological health. Companies across sectors—from manufacturing to technology—adopt closed-loop systems that minimize waste, reuse materials, and reduce pollution. The principles of circular economy take root, transforming production processes to align with the Earth's natural cycles. Waste is no longer an inevitable byproduct of industry but a resource that can be repurposed and reintegrated. In this way, energy abundance becomes a catalyst for a broader transformation, prompting humanity to rethink its impact on the planet and its role as a steward of Earth's resources.

With energy no longer a scarce commodity, society is freed from the pressures that have driven conflict and inequality. The pursuit of power shifts from a struggle over limited resources to a shared commitment to sustainable growth. Nations find new opportunities for collaboration, recognizing that their shared interest in a stable climate and healthy ecosystems outweighs any historical rivalries. This spirit of cooperation enables countries to pool their knowledge, technology, and resources, building a resilient global infrastructure that supports energy access, environmental health, and economic stability. International partnerships emerge, not out of necessity but out of mutual respect and a collective dedication to ensuring that the benefits of energy abundance are equitably distributed.

This new model of international cooperation is characterized by transparency, fairness, and shared accountability. Nations work together to establish standards that protect the environment, support ethical energy practices, and prevent the monopolization of resources. Global institutions play a crucial role, providing oversight and fostering a collaborative spirit that unites countries in the pursuit of shared goals. These institutions act as custodians, ensuring that the infrastructure of energy abundance remains accessible to all and that no single entity holds undue influence. This framework of governance reflects a profound shift in global relations, where the principles of equity and inclusivity take precedence over competition and self-interest.

The cultural shift toward sustainability and responsibility is not limited to governments and industries; it extends into the hearts and minds of individuals, shaping everyday choices and values. Energy abundance inspires a lifestyle that values mindfulness, conservation, and intentionality, fostering a society where people are empowered to make choices that benefit both themselves and the planet. This shift is reflected in the design of homes, which incorporate energy-efficient appliances, sustainable materials, and renewable power sources. Individuals are encouraged to adopt practices that reduce waste, conserve water, and support local food systems, creating a culture of sustainability that permeates every aspect of life.

Education becomes a key component of this cultural transformation, equipping individuals with the knowledge and skills needed to thrive in an energy-abundant world. Schools and universities prioritize environmental literacy, ethical reasoning, and critical thinking, preparing students to navigate the complexities of a sustainable society. Curricula emphasize the interconnectedness of human systems and natural ecosystems, fostering an awareness of the impact that each person has on the world around them. By instilling these values from a young age, education cultivates a generation of informed, responsible citizens who are prepared to carry forward the legacy of energy abundance with integrity and compassion.

In this world, the arts flourish as a means of expression, connection, and reflection. Artists, inspired by the values of sustainability and unity, create works that celebrate humanity's relationship with the Earth and explore the ethical dimensions of technological progress. Art becomes a powerful tool for fostering empathy and understanding, bridging cultural divides and reminding people of their shared humanity. Public art installations, community festivals, and cultural events bring people together, creating spaces for dialogue, celebration, and introspection. Through the arts, society finds a way to explore the complexities of this new era, to honor its triumphs and confront its challenges, and to celebrate the resilience and beauty of the human spirit.

As humanity adapts to the realities of energy abundance, a new relationship with technology emerges. No longer a driver of resource depletion, technology is reimagined as a force for conservation and regeneration. Innovations in artificial intelligence, robotics, and biotechnology are guided by ethical considerations, prioritizing applications that enhance quality of life while respecting the environment. Engineers and scientists work to develop solutions that not only meet human needs but do so in ways that are sustainable and equitable. Technology becomes a partner in humanity's journey toward balance, a tool for nurturing life rather than consuming it.

The role of artificial intelligence in this new world is particularly significant. AI, with its capacity to process vast amounts of data and solve complex problems, becomes an essential tool in managing the infrastructure of energy abundance. AI systems monitor energy flows, optimize resource distribution, and predict environmental impacts, ensuring that the benefits of renewable power are maximized and that potential risks are mitigated. These systems operate transparently and ethically, guided by principles that prioritize human welfare and environmental integrity. In this context, AI is not an autonomous force but an extension of humanity's values, a testament to the power of technology when aligned with ethical purpose.

The transformative potential of energy abundance also reshapes the landscape of healthcare, empowering communities to achieve unprecedented levels of well-being. With reliable access to energy, healthcare facilities are equipped with advanced medical technologies that support both preventive and responsive care. Remote areas gain access to telemedicine services, bridging the gap between urban and rural health resources and ensuring that every individual has the opportunity to receive timely, high-quality care. Hospitals and clinics operate sustainably, using renewable energy to power life-saving equipment and maintain essential services. This comprehensive healthcare infrastructure extends life expectancy and enhances quality of life, creating a society where health and wellness are not privileges but universal rights.

In this new era, humanity's relationship with the environment becomes one of mutual respect and cooperation. Freed from the pressures of resource extraction, people can focus on restoration, conservation, and responsible stewardship. Large-scale projects aimed at reforestation, soil regeneration, and biodiversity protection thrive, supported by the abundance of clean energy that enables these initiatives without placing additional strain on natural systems. Oceans, forests, and wetlands—once threatened by pollution and exploitation—are restored to health, creating vibrant ecosystems that sustain a diversity of life. This commitment to ecological integrity reflects humanity's recognition of the Earth not as a resource to be controlled but as a partner in a shared journey.

The principles of energy abundance inspire a cultural renaissance, where progress is defined not by the pace of development but by its depth and purpose. Society embraces a slower, more intentional approach to growth, where the focus shifts from material accumulation to the enrichment of human experience. This renaissance is characterized by a return to values of balance, compassion, and respect, a collective affirmation that true progress is achieved when humanity lives in harmony with the natural world. Communities come together to celebrate this shift, recognizing that energy abundance is not an endpoint but a new beginning—a foundation upon which to build a society that honors the interconnectedness of all life.

As humanity steps into this future, it carries with it the lessons of the past, the awareness that power, when misused, can lead to division and destruction. But in the hands of an enlightened society, energy abundance becomes a force for unity and renewal. It offers a chance to redefine success, to reshape the fabric of civilization, and to create a legacy that reflects the highest ideals of humanity. This vision is not a utopian dream but a practical reality, one that requires commitment, cooperation, and the courage to embrace a new way of living.

In this era, energy abundance becomes a beacon, illuminating the path toward a world where peace, sustainability, and shared prosperity are not ideals but everyday realities. It serves as a reminder that humanity's greatest achievements lie not in the conquest of nature but in the wisdom to live within its bounds, to respect its rhythms, and to protect its beauty. The journey toward energy abundance is a journey of awakening, a call to recognize that the Earth, with all its wonders, is not a possession but a responsibility, a gift to be cherished and preserved for generations to come.

In this way, energy abundance does more than power machines or fuel industries—it powers hope, resilience, and the endless capacity of the human spirit to adapt, to grow, and to find meaning. It is a legacy that reaches beyond technology, beyond economics, and beyond politics, touching the very essence of what it means to be human. In choosing this path, humanity chooses to build a world where every life is valued, where every ecosystem is respected, and where the pursuit of knowledge, beauty, and justice is boundless and free.

As humanity embraces the possibilities of an energy-abundant world, the vision of progress takes on new depth and dimension. In this era, technological achievements are woven into a framework that values compassion, responsibility, and inclusivity, a framework that transforms the role of power from a tool of division into a force of unity. Energy abundance, once an aspiration, now serves as a foundation upon which societies can build resilient infrastructures, equitable economies, and cultures that thrive in harmony with the natural world. This era does not merely extend the limits of human potential; it redefines those limits, aligning them with principles that honor both human dignity and environmental integrity.

With the abundance of clean, renewable energy, the possibilities for scientific and technological advancement expand exponentially. Freed from the constraints of limited resources, researchers explore the frontiers of knowledge, pushing the boundaries of what is possible in fields such as medicine, environmental science, and space exploration. Laboratories, powered by sustainable energy, delve into the mysteries of the human genome, exploring therapies and innovations that address diseases previously thought incurable. This biomedical renaissance is not confined to the privileged few but is shared openly, driven by a commitment to global health and the understanding that wellness, like energy, must be universally accessible.

Environmental science, too, is redefined in this new age, as energy abundance enables the development of technologies that restore and protect the planet. Sophisticated sensors, AI-driven analytics, and real-time environmental monitoring systems provide unprecedented insight into the health of ecosystems, allowing scientists to identify and address environmental threats with precision. These technologies, powered by renewable energy, support large-scale projects in reforestation, soil regeneration, and ocean conservation, healing landscapes that have borne the scars of industrial expansion. The abundance of energy empowers humanity to restore balance, transforming previously damaged environments into thriving ecosystems that support biodiversity and resilience.

In the vast expanse of space, humanity's pursuit of knowledge reaches new heights. The abundant power of renewable energy allows for the construction and maintenance of orbital research stations, deep-space probes, and even planetary habitats, advancing exploration beyond the limits of Earth. Space exploration is no longer an endeavor of geopolitical competition but a shared journey that unites nations in the pursuit of universal understanding. Scientists, astronauts, and engineers work collaboratively, fueled by a vision that sees humanity as part of a greater cosmic family. This spirit of exploration does not seek to conquer space but to understand it, to embrace the mysteries of the universe with humility and awe. Each discovery deepens humanity's sense of wonder, reinforcing the belief that knowledge is a gift to be shared, a bridge that connects people across cultures, generations, and worlds.

Back on Earth, the abundance of energy transforms communities and everyday life, creating a foundation for equitable access to resources and opportunities. Housing is reimagined to be sustainable, comfortable, and accessible, with homes powered by renewable energy and designed to coexist seamlessly with their environments. Buildings are constructed with materials that are renewable, recycled, and locally sourced, reducing environmental impact while fostering a sense of place and connection to the land. Cities incorporate green spaces, renewable energy infrastructure, and public art, creating environments that inspire well-being and creativity. These communities are not designed merely to function but to flourish, embodying a vision of urban life that balances the practical needs of residents with the beauty and tranquility of natural landscapes.

In this energy-abundant world, transportation becomes both accessible and environmentally harmonious. Public transit systems powered by clean energy connect cities, rural areas, and remote regions, ensuring that people can travel freely without compromising the health of the planet. High-speed trains, electric buses, and sustainable urban transport systems replace fossil-fuel-dependent vehicles, creating a transportation network that is efficient, affordable, and low-impact. These innovations enable people to move seamlessly within and between communities, reducing the need for personal vehicles and contributing to a cleaner, quieter, and more connected world. The new paradigm of transportation is not only a means of mobility but a reflection of society's commitment to sustainability and inclusivity, ensuring that every individual, regardless of location or economic status, can access the opportunities and connections they need to thrive.

The transformative impact of energy abundance extends to global economies, reshaping markets and industries with an emphasis on fairness, sustainability, and resilience. With energy no longer a scarce resource, economic models shift away from competition over limited supplies toward collaboration, innovation, and regenerative practices. Local economies flourish as communities produce their own renewable power, creating self-sustaining systems that reduce dependency on external sources and empower individuals to contribute to their local energy networks. In this economy, prosperity is measured not by consumption but by the capacity to regenerate resources, reduce waste, and support the well-being of people and ecosystems alike. Industries that once thrived on extraction adapt to a circular model, where products are designed to last, to be reused, and to reintegrate into the environment without harm.

In education, the principles of energy abundance foster a culture of lifelong learning and global exchange. Schools and universities, supported by sustainable energy, provide students with access to digital resources, virtual classrooms, and research tools that connect them to a global community of learners. This educational transformation enables students to develop skills and knowledge in fields that are not only practical but ethically grounded, emphasizing critical thinking, environmental awareness, and social responsibility. Education becomes a pathway not just to individual achievement but to collective progress, preparing students to become leaders in a society that values sustainability, inclusivity, and innovation. With barriers to access removed, knowledge becomes a shared asset, empowering individuals from all walks of life to contribute to the global conversation and advance solutions for the challenges of the future.

The arts and humanities also flourish, reflecting and shaping the values of an energy-abundant society. Artists, writers, musicians, and thinkers explore themes of environmental stewardship, human connection, and ethical responsibility, creating works that inspire reflection, dialogue, and action. Art becomes a medium through which society explores its evolving relationship with nature, technology, and each other, serving as a reminder of the beauty and diversity that define human experience. Public spaces, cultural institutions, and community festivals celebrate creativity and encourage participation, fostering a sense of unity and shared identity. In this way, the arts serve as both a mirror and a guide, helping society navigate the complexities of energy abundance and the responsibilities that come with it.

As the era of energy abundance continues to shape humanity's future, the principles of fairness, transparency, and collaboration become embedded in the governance of energy resources. Governments, guided by the needs of their citizens and the health of the planet, establish frameworks that ensure equitable access, protect against exploitation, and promote responsible use. International coalitions emerge to uphold these principles, fostering a global infrastructure that prioritizes the well-being of all people and ecosystems. This governance model reflects a profound shift in the relationship between power and accountability, recognizing that energy abundance is not merely a resource to be managed but a trust to be upheld.

The role of global institutions in this framework is to act as stewards of energy abundance, ensuring that policies reflect the highest ethical standards and that resources are managed in a way that benefits all. These institutions work in partnership with local governments, private sectors, and civil society, creating a network of accountability and support that reinforces the values of sustainability and equity. By fostering a collaborative approach, these institutions help bridge gaps between nations, ensuring that energy abundance does not become a tool of dominance but a means of fostering peace, stability, and shared progress.

This model of governance is built on the belief that energy, like air and water, is a universal right. It reflects a commitment to inclusivity and a recognition that true abundance can only be achieved when every individual has access to the resources they need to lead a healthy, meaningful life. In this vision, energy abundance is not an endpoint but a foundation, a resource that supports a just, resilient, and compassionate society. It is a commitment to future generations, a promise that the benefits of today's innovations will be preserved, protected, and shared.

This promise, embodied in the very structure of society, reflects humanity's evolution from a species driven by survival to one guided by purpose. Energy abundance is the culmination of centuries of scientific inquiry, ethical reflection, and social progress, an achievement that symbolizes the highest aspirations of human civilization. It is a testament to the power of unity, a reminder that when people come together with a shared vision, they can create a world that honors the dignity of every life and the beauty of the natural world.

As humanity navigates this new era, it carries with it a deepened sense of responsibility. The abundance of energy invites reflection on the kind of world that each generation wishes to build, a world where resources are not fought over but cherished, where progress is not measured by dominance but by harmony. This reflection guides society's choices, encouraging leaders, innovators, and citizens to prioritize actions that protect the environment, uplift communities, and create a legacy of sustainability. In this way, energy abundance becomes more than a source of power; it becomes a moral compass, a guiding light that leads humanity toward a future of shared prosperity and peace.

The journey to energy abundance reveals the potential for humanity to transcend the limitations of the past, to build a civilization that is resilient, inclusive, and deeply connected to the Earth. This journey is not without challenges, but each step forward brings with it new insights, new opportunities, and new reasons to hope. Humanity's embrace of renewable energy marks a turning point, a commitment to forge a path that honors life in all its forms and respects the planet that sustains it.

In the end, energy abundance is a story of renewal—a story that redefines what it means to thrive as a society, to progress as a civilization, and to coexist as part of a larger ecosystem. It is a story that celebrates the ingenuity, compassion, and resilience of humanity, a reminder that the journey toward a sustainable future is a journey worth taking. In choosing this path, humanity commits not only to preserving the planet but to nurturing the potential within each person, to creating a world where every life is valued, every ecosystem is protected, and every step forward is a step toward a more just, harmonious, and enlightened existence.

As the final echoes of energy scarcity fade, the vision of energy abundance stands as a beacon—a symbol of humanity's capacity for transformation, a testament to the enduring power of hope, and a promise that the future can be as boundless as the energy that fuels it.

As humanity fully embraces the profound implications of energy abundance, the very essence of civilization is transformed. The dawn of this new era is marked by a collective realization that progress and prosperity need not come at the expense of the environment, nor should they be limited to select groups or regions. Instead, energy abundance brings with it a commitment to universal equity, environmental stewardship, and a shared purpose that transcends cultural, national, and economic boundaries. In this world, energy is not a source of competition but a foundation upon which societies build a sustainable and just future, weaving technology, ethics, and human potential into a seamless narrative of shared growth.

With energy abundance, the concept of sustainability is no longer confined to the margins of policy or public consciousness; it becomes the heart of every decision, every innovation, and every aspect of daily life. The global community, once fragmented by conflicting interests and resource disparities, is united by a common commitment to honor the planet and its finite resources. This era marks the end of exploitation and the beginning of regeneration, where humanity steps into its role not as a conqueror of nature but as a careful steward, respecting the delicate balance that sustains life on Earth. As ecosystems are restored, biodiversity flourishes, and the natural world reclaims its resilience, humanity learns to live within the boundaries set by nature, to thrive not despite these limits but because of them.

In this transformed society, the principles of energy abundance foster not only technological advancement but also spiritual and philosophical growth. Humanity's journey to energy abundance is as much an internal evolution as it is an external achievement, prompting people to reflect on the values that define a meaningful life. Freed from the constant pressures of scarcity, individuals and communities are encouraged to explore deeper questions of purpose, connection, and legacy. What does it mean to thrive? How can one contribute to a world that honors the dignity of all life? In answering these questions, people are drawn to values of empathy, mindfulness, and unity, creating a cultural ethos that celebrates both individual fulfillment and collective well-being.

This cultural shift toward ethical living and shared purpose permeates every layer of society, influencing art, education, governance, and personal relationships. Artists, inspired by a world where energy is abundant and creativity is celebrated, produce works that explore themes of interconnectedness, environmental harmony, and social justice. Art becomes a means of fostering empathy, bridging divides, and inspiring people to envision a future defined not by material wealth but by the richness of human experience. Public art installations, murals, and sculptures transform urban landscapes into spaces of reflection and community, reminding people of their shared responsibility to the Earth and to each other. Cultural festivals, performances, and exhibitions celebrate the diversity of human expression and the unity of human purpose, creating moments of collective joy and introspection.

In education, the legacy of energy abundance is realized through a commitment to holistic learning that prepares individuals not only for professional success but for responsible citizenship in a sustainable world. Schools, colleges, and research institutions place a strong emphasis on environmental literacy, ethics, and global citizenship, nurturing a generation of leaders who are equipped to tackle the complexities of an interconnected world. Students learn to think critically, to question, and to innovate, understanding that knowledge is a tool for positive change. This new model of education fosters an awareness of the individual's impact on society and the environment, inspiring a lifelong dedication to stewardship and service. By instilling these values, education becomes a pathway not only to personal achievement but to collective progress, a promise that each generation will carry forward the principles of equity and respect that define the age of energy abundance.

As energy abundance shapes a more inclusive and responsible society, governance also evolves to reflect these values. In this era, governments are guided not by the imperatives of growth and control but by the principles of fairness, transparency, and sustainability. Public policy is crafted with an awareness of the long-term impact on both people and the planet, ensuring that decisions reflect the highest ethical standards and the interests of future generations. Leaders are chosen not for their political might but for their commitment to service, their dedication to ethical governance, and their vision for a society that thrives in harmony with the natural world. International alliances are built on mutual respect, with nations working together to uphold the principles of peace, equity, and environmental protection. This collaborative approach to governance embodies the belief that global challenges require global solutions and that a sustainable future can only be achieved through unity and cooperation.

The principles of governance in this world of abundance extend beyond national borders, fostering a global community that recognizes its shared responsibility to steward the Earth. International organizations serve as guardians of energy equity, ensuring that all countries and communities benefit from the advancements made possible by renewable power. These institutions act as arbiters of fairness, facilitating access to energy resources, supporting sustainable development, and preventing monopolization. In this way, governance becomes a testament to humanity's capacity for ethical leadership, a reflection of the belief that power is best exercised when it serves the greater good.

The relationship between humanity and technology, too, is redefined in this era. With energy no longer a limiting factor, technology becomes a partner in the journey toward a sustainable, compassionate world. Innovations in artificial intelligence, biotechnology, and environmental engineering are developed and deployed with ethical considerations at the forefront, prioritizing applications that enhance quality of life while safeguarding the environment. AI-driven systems support everything from energy distribution and waste management to healthcare and climate resilience, ensuring that resources are used efficiently and equitably. These technologies, guided by principles of transparency and accountability, operate as extensions of humanity's values, a testament to the power of innovation when aligned with ethical purpose. In this context, technology is no longer a driver of resource consumption but a tool for nurturing life, supporting ecosystems, and building a legacy of balance and respect.

This profound alignment between technology and ethics is perhaps most visible in the field of healthcare, where energy abundance transforms the way societies care for their members. With reliable, clean power available in every corner of the globe, healthcare facilities operate at peak capacity, delivering consistent, high-quality care to all. Telemedicine becomes a standard service, bridging the gap between rural and urban healthcare resources and ensuring that distance is no barrier to access. Hospitals, clinics, and research centers are powered sustainably, allowing medical professionals to focus not on energy costs but on patient outcomes and preventative care. Medical research flourishes, driven by the understanding that health is a universal right and that every life is valuable. This comprehensive healthcare infrastructure creates a world where wellness is within reach for all, where each individual is supported in leading a healthy, fulfilling life.

As humanity embraces the benefits of energy abundance, a new relationship with the environment emerges—one of respect, gratitude, and active care. The natural world, once viewed primarily as a resource for extraction, is now recognized as a partner in the quest for sustainability. Conservation efforts flourish, supported by renewable energy that enables large-scale environmental projects without placing additional strain on ecosystems. Forests are restored, wetlands are preserved, and oceans are protected, creating vibrant habitats that sustain diverse forms of life. Humanity's footprint, once heavy and harmful, becomes light and regenerative, leaving the Earth healthier and more resilient. This commitment to ecological stewardship reflects humanity's understanding that true abundance is achieved only when the planet itself thrives.

The legacy of energy abundance, then, is more than technological advancement; it is a transformation in how humanity views itself and its place in the world. It is a shift from a mindset of dominance to one of partnership, a recognition that the well-being of people and the planet are inseparable. This legacy is a promise to future generations, a commitment to leave a world that is as beautiful, balanced, and bountiful as it was found. Energy abundance is the foundation of this promise, a resource that fuels not only machines and economies but the very ideals of fairness, compassion, and unity.

In this vision of the future, energy abundance serves as a catalyst for a new kind of human progress, one that is measured not by material wealth but by the richness of human experience, the health of the planet, and the strength of community bonds. This progress is reflected in the choices made by individuals, the policies enacted by governments, and the innovations developed by scientists and engineers. It is a progress that honors the interconnectedness of all life, a progress that celebrates diversity, resilience, and the boundless potential of human creativity.

As humanity steps into this future, it carries with it a sense of gratitude, an appreciation for the journey that has brought it to this point and a respect for the responsibility that comes with abundance. The story of energy abundance is not one of conquest or competition; it is a story of renewal, a testament to the power of collective vision and shared purpose. It is a reminder that the greatest achievements are those that lift others up, that protect and preserve, that inspire and unite.

In choosing this path, humanity embraces the possibility of a world where energy flows freely, where resources are shared equitably, and where progress is driven by a commitment to peace, justice, and sustainability. This world is not a distant dream but a practical reality, built on the foundation of science, ethics, and human resilience. It is a world where every life is valued, where every voice is heard, and where every step forward is a step toward a more harmonious, compassionate, and enlightened existence.

Energy abundance, in this vision, is more than a source of power; it is a symbol of humanity's highest ideals, a beacon that illuminates the path to a future where the potential for growth, discovery, and joy is as boundless as the energy that fuels it. This is the legacy of energy abundance—a legacy of hope, resilience, and the enduring power of humanity to create a world that reflects the best of what it means to be human.

Chapter 2: Zero-Gravity and Superconductivity—The Physics Behind the Vision

In the pursuit of energy abundance, humanity's journey is not just one of technological progress; it is a transformation that redefines the boundaries of science, society, and the human spirit. The ambition to harness limitless power, once the domain of dreams and speculation, now drives our most audacious explorations and innovations. This endeavor is more than the quest to end energy scarcity—it is a mission to reshape civilization's relationship with the natural world, to create systems that serve not only present needs but also the well-being of future generations. In this way, the journey toward energy abundance is as much an ethical commitment as it is a scientific pursuit, calling for a vision that bridges knowledge, responsibility, and courage.

The foundation of this vision lies in the very fabric of physics, in the forces that bind matter and govern the universe. As humanity's understanding of these principles deepens, so too does the potential to leverage them for the benefit of all. Central to this exploration is the concept of energy itself—an elusive force that can neither be created nor destroyed, only transformed. This principle, articulated in the first law of thermodynamics, has guided physicists and engineers alike, shaping our approach to energy generation, storage, and consumption. However, as we step closer to an age defined by abundance, these principles invite new interpretations, challenging humanity to rethink the ways energy is captured, stored, and used.

The science of energy transformation reaches back to the discovery of steam power, when innovators learned to convert heat into mechanical motion. With each subsequent leap in technology, humanity came closer to understanding the true potential of energy. Yet, with each advancement came a deeper realization of the limitations inherent in existing methods. Coal and oil, while transformative, bound society to a finite, pollutive cycle. The early 20th century's development of nuclear fission offered a glimpse of boundless energy, but with it came risks and challenges that underscored the need for caution and respect for the immense forces we sought to harness. Each breakthrough carried humanity forward, yet also imparted a lesson in humility, reinforcing the need for an approach that balances power with prudence.

As the 21st century unfolds, the vision of energy abundance embraces a new generation of technologies that seek to transcend these limitations. Scientists are exploring renewable sources—solar, wind, geothermal, and tidal energy—not as stopgaps but as the foundation of a new energy paradigm. These sources, derived from natural processes, present a way of capturing energy without depleting resources or polluting ecosystems. Yet they also introduce new challenges, for while the sun, the wind, and the Earth's heat are inexhaustible, their availability fluctuates. The sun does not always shine; the wind does not always blow; the Earth's internal heat is not evenly distributed. Harnessing these forces requires ingenuity, a commitment to overcoming the inherent intermittency and regional disparities that characterize renewable energy sources.

At the heart of these innovations lies the concept of energy storage, an arena of science and engineering that has seen exponential growth as society seeks ways to stabilize and distribute power. Energy storage systems are more than mere batteries; they are complex, dynamic solutions that transform renewable energy from a variable asset into a reliable resource. Advanced lithium-ion and solid-state batteries have opened new possibilities, enabling the storage of vast amounts of power in compact, efficient forms. Beyond these, researchers are pioneering the development of flow batteries, hydrogen fuel cells, and compressed air systems, each representing a unique approach to energy storage. These technologies are not only tools; they are stepping stones in humanity's journey toward independence from fossil fuels and toward the promise of a world powered by nature's most enduring forces.

The pursuit of effective storage solutions has led scientists to reimagine the very structure of the electrical grid, envisioning networks that are as flexible as they are resilient. Traditional grids, built to distribute energy from centralized fossil-fuel power plants, lack the adaptability needed for an era of renewable energy. In response, engineers are designing smart grids—networks that dynamically balance supply and demand, integrating solar panels, wind turbines, and battery systems across vast areas. Smart grids are capable of responding in real-time, directing energy where it is needed most, optimizing efficiency, and minimizing waste. By connecting decentralized sources of power, smart grids transform the grid from a one-way flow of electricity into a complex, interactive ecosystem, where energy flows in multiple directions and is accessible to all.

Yet even as these advancements bring us closer to a sustainable future, they reveal the necessity of an approach that is not only technologically sophisticated but philosophically grounded. The journey to energy abundance is fraught with ethical questions that demand consideration. How can we ensure that the benefits of renewable energy are shared equitably, reaching communities and nations that have historically been left behind in the race for progress? How can we balance the drive for innovation with the need to protect the delicate ecosystems upon which life depends? These questions are not tangential to the pursuit of energy abundance; they are central to it, guiding the choices we make as individuals, as societies, and as a global civilization.

The movement toward energy abundance is, therefore, as much a moral endeavor as it is a technological one. The vision of a world powered by renewable energy must be inclusive, ensuring that access to power is not dictated by wealth or geography but is viewed as a universal right. This requires a commitment to global cooperation, as nations work together to share knowledge, resources, and technological breakthroughs.

International alliances and agreements become essential, providing a framework within which energy can be managed as a shared resource, rather than a commodity to be controlled. The ethical framework of energy abundance thus transcends borders, creating a foundation for peace, equity, and environmental stewardship.

In this light, the pursuit of energy abundance calls for a reexamination of the values that drive society. The relentless pursuit of growth and consumption, once seen as the hallmark of progress, is no longer compatible with a world that values sustainability. Instead, humanity must embrace a philosophy of regeneration, prioritizing practices that replenish rather than exhaust, that build rather than break down. This regenerative approach extends beyond energy to encompass agriculture, industry, and urban planning, fostering systems that operate in harmony with the environment rather than in opposition to it. By aligning human activities with the cycles of nature, society creates a model of development that is as resilient as it is sustainable.

As humanity stands on the threshold of this new era, the path forward is illuminated by the lessons of the past. Each step in the journey toward energy abundance has underscored the need for a balanced approach, one that recognizes the power of technology while respecting the limits of nature. The story of energy, from steam engines to smart grids, is a testament to the ingenuity and adaptability of the human spirit, a reflection of the capacity to innovate and evolve. But it is also a reminder of the need for humility, for an awareness that the forces we seek to harness are both awe-inspiring and unforgiving. As we pursue the promise of abundance, we are called to remember that true progress is not measured by the power we command but by the integrity with which we wield it.

In this journey, humanity finds itself not merely on a path to technological advancement but on a quest for balance—a quest that requires a reevaluation of what it means to be prosperous, to be progressive, and to be human. The dream of energy abundance is, at its core, a dream of harmony: harmony between people and nature, between innovation and preservation, between the needs of today and the rights of tomorrow. It is a vision that calls upon the deepest reserves of wisdom, compassion, and courage, challenging society to create a world where energy flows not only through wires and grids but through the ideals that unite and uplift.

The pursuit of energy abundance is thus a journey of interconnectedness, one that demands a collaborative spirit and a commitment to the greater good. The technology that makes this vision possible is, in many ways, secondary to the spirit that drives it—the determination to build a world that is just, sustainable, and inclusive. In this way, the journey to energy abundance is not simply a scientific endeavor but a social revolution, a reimagining of civilization itself.

The promise of energy abundance opens the door to an era of innovation where science and technology are mobilized not only to sustain life but to enhance it. Each breakthrough in renewable energy, each advancement in storage, and each refinement of the smart grid system represents a step toward a world where energy is not only plentiful but a force for equity and environmental harmony. But as humanity advances, the quest for sustainable energy becomes as much about the architecture of society as it is about the mechanics of power generation. Energy abundance invites humanity to reimagine the infrastructure of civilization itself, to rebuild cities, industries, and communities around principles that support regeneration and resilience.

Central to this vision are the smart grid systems that transform the way energy is generated, distributed, and used. Unlike traditional grids, which depend on centralized power plants and rigid distribution networks, smart grids operate as dynamic, adaptive systems capable of responding to fluctuations in supply and demand. Through a complex interplay of algorithms, sensors, and artificial intelligence, smart grids enable decentralized power generation, drawing from solar panels on rooftops, wind turbines in rural areas, and community battery systems. This decentralization democratizes energy access, allowing communities to generate and share power autonomously, fostering resilience and reducing reliance on centralized, often fossil-fuel-dependent, power sources.

These smart grids embody a vision of interconnectivity, where energy flows seamlessly across borders and communities, adapting to shifts in weather patterns, consumption habits, and technological changes. By integrating renewable sources and battery storage across vast regions, smart grids can deliver reliable power regardless of individual source variability. Solar farms in sunny regions complement wind farms in windy locales, while energy storage facilities smooth out peaks and troughs in supply. The grid itself becomes a living entity, constantly adjusting and reallocating resources, ensuring that no community is left in the dark and no surplus goes to waste. Through this interconnected network, energy abundance begins to reshape not only the power infrastructure but the social contract, creating a foundation for a society that values collaboration, sustainability, and inclusivity.

The emergence of smart grids also highlights the role of data and artificial intelligence in managing the complexities of renewable energy. Real-time data from sensors, weather forecasting models, and consumption patterns feeds into AI-driven systems that optimize the flow of energy across the grid. These systems learn from patterns, anticipate demand, and adjust supply accordingly, maximizing efficiency and minimizing waste. AI's capacity to analyze and respond to vast amounts of data allows the grid to operate with precision, directing power to areas of highest need and reducing the strain on any single source. Through these intelligent systems, humanity is able to harness renewable energy at a scale and efficiency once unimaginable, transforming potential intermittencies into opportunities for balance and synergy.

However, as the energy grid becomes increasingly intelligent and adaptive, questions arise about governance, privacy, and accountability. The use of AI to manage energy flows introduces ethical considerations that extend beyond the technical domain. Who owns the data that drives the grid? How can society ensure that these systems operate transparently and equitably, serving the interests of the public rather than private entities? The answers to these questions will shape the relationship between technology and society in the age of energy abundance, underscoring the need for frameworks that prioritize transparency, protect individual rights, and maintain public trust.

The pursuit of energy abundance also reshapes the physical and social landscapes of urban and rural areas alike. As smart grids become the norm, cities are redesigned to integrate renewable energy sources at every level, from public infrastructure to private residences. Buildings are equipped with solar panels, green roofs, and energy-efficient systems, transforming them into power generators as well as consumers. Homes are built to be self-sustaining, with energy storage solutions that allow residents to produce, store, and share power within their communities. This new model of urban design fosters a culture of sustainability, where energy production is woven into the fabric of daily life, and where people are empowered to contribute actively to the collective energy ecosystem.

These changes extend beyond the boundaries of cities, bringing new opportunities for rural areas as well. In regions once marginalized by a lack of infrastructure, energy abundance enables the development of localized power networks that connect rural communities to the broader grid while maintaining autonomy. These microgrids, powered by solar, wind, and geothermal energy, provide a reliable source of power that supports agriculture, education, and healthcare. Rural communities become self-sufficient, with the ability to generate their own power, fostering economic resilience and reducing dependency on external resources. In this model, energy abundance empowers people in all corners of society, ensuring that access to energy is not limited by geography or socioeconomic status but is recognized as a universal right.

As cities and communities evolve to reflect these values, the concept of urban resilience takes on new meaning. Energy abundance creates the foundation for cities that are not only sustainable but adaptable, able to withstand and recover from the challenges posed by climate change and other environmental stresses. Buildings are designed to reduce energy consumption and minimize environmental impact, using materials that are renewable, recycled, and locally sourced. Public transportation systems, powered by clean energy, reduce pollution and provide efficient alternatives to private vehicles. Green spaces, urban gardens, and natural water management systems become integral parts of city planning, creating environments that support both biodiversity and human well-being. These resilient cities are designed not only to meet the needs of their residents but to coexist harmoniously with the ecosystems that surround them, fostering a culture that values balance, diversity, and coexistence.

The transformation of urban and rural landscapes in the age of energy abundance also fosters a renewed commitment to ecological integrity. Freed from the need to extract and consume finite resources, humanity is able to shift its focus from exploitation to preservation and restoration. Large-scale reforestation projects, soil regeneration initiatives, and habitat conservation efforts flourish, supported by renewable energy that powers these undertakings without depleting natural resources. This commitment to ecological health reflects a profound shift in the values that underpin society, a recognition that the well-being of humanity is inseparably linked to the health of the planet. As energy abundance enables the healing of ecosystems, it also fosters a mindset of respect and responsibility, encouraging people to view the natural world not as a resource to be consumed but as a legacy to be protected.

The economic implications of energy abundance extend deeply into every industry, reshaping traditional models of production, consumption, and distribution. In a world where energy is abundant, businesses no longer compete over limited resources but collaborate to create sustainable systems that benefit all.

Manufacturing processes are redesigned to minimize waste, reduce carbon emissions, and promote circularity, where materials are reused, recycled, and reintegrated into the production cycle. This circular economy transforms industries from extractive enterprises into regenerative ones, where the focus shifts from short-term profits to long-term sustainability. Companies that once relied on fossil fuels to power their operations now invest in renewable energy, aligning their business models with the principles of environmental stewardship and social responsibility.

This transformation in industry is not limited to manufacturing; it extends to every sector, from agriculture to healthcare, from transportation to finance. In agriculture, renewable energy supports innovations such as vertical farming, precision agriculture, and sustainable irrigation, enabling food production that is efficient, resilient, and ecologically sound. In healthcare, energy abundance powers hospitals, clinics, and research facilities in even the most remote areas, ensuring that everyone has access to quality care. Transportation networks, powered by renewable energy, connect people and communities in ways that are affordable, accessible, and environmentally friendly. Financial institutions, guided by principles of sustainability, invest in projects that promote renewable energy, green infrastructure, and social equity. This holistic transformation creates an economy that values human and environmental well-being as much as economic growth, fostering a society where prosperity is measured not by consumption but by the capacity to regenerate and sustain.

As energy abundance reshapes the economic landscape, it also challenges individuals and communities to rethink their relationship with resources, consumption, and progress. The relentless drive for accumulation and expansion, once seen as synonymous with success, is replaced by a commitment to balance, mindfulness, and intentionality. This cultural shift reflects a deeper understanding of what it means to thrive, a recognition that true prosperity is found not in the endless pursuit of more but in the thoughtful stewardship of what is available. In this way, energy abundance fosters a culture that values quality over quantity, resilience over rapid growth, and long-term well-being over immediate gains.

This transformation of values permeates every aspect of life, shaping the choices people make as consumers, the policies they support as citizens, and the innovations they pursue as creators. As energy abundance makes sustainability accessible, individuals are empowered to live in alignment with their values, making choices that contribute to a healthier planet and a more just society. Homes are designed to be energy-efficient, with appliances and systems that reduce waste and conserve resources. Communities come together to support local food systems, renewable energy initiatives, and public spaces that foster connection and environmental awareness. This new lifestyle is not driven by sacrifice or scarcity but by a sense of purpose and fulfillment, a recognition that living sustainably is a way of honoring both the Earth and future generations.

As society embraces these principles, education becomes a key avenue for instilling the values of energy abundance in future generations. Schools, universities, and community programs emphasize environmental literacy, ethical reasoning, and global citizenship, preparing students to navigate the complexities of a world that values sustainability and equity. Education is no longer focused solely on preparing students for the workforce; it is a journey of personal and social growth, where students are encouraged to explore their roles as stewards of the planet and as members of a global community. This approach to education fosters a generation that is not only knowledgeable but also compassionate, resilient, and committed to building a world that reflects the ideals of energy abundance.

As the vision of energy abundance unfolds, society witnesses a profound transformation in the principles that guide growth, innovation, and human connection. Each advancement is not just a technical achievement but a step toward redefining the ideals of prosperity and progress, fostering a world where the well-being of people and the planet are held in harmony. This new paradigm shifts the focus from individual gain to collective resilience, from unchecked consumption to thoughtful regeneration. As energy becomes a universally accessible resource, it empowers societies to flourish without compromising the ecological balance that sustains all life.

The transformation enabled by energy abundance brings a reimagining of industries and sectors that once relied heavily on finite resources. With renewable power driving every aspect of production, industries become symbols of sustainability rather than exploitation. Manufacturing evolves into a circular process, where every material, every component is designed with a lifecycle that respects the Earth's finite resources. Factories that once emitted pollutants now operate with near-zero waste, utilizing advanced recycling methods and sustainable materials. Products are crafted not just to serve consumers but to return to the ecosystem without leaving a harmful footprint. This shift in manufacturing priorities is a testament to humanity's commitment to integrating technology and ethics, an alignment that reflects a broader cultural dedication to respect the environment.

In agriculture, energy abundance fosters innovations that meet humanity's growing needs without exhausting the land and water resources on which farming depends. Energy-efficient systems like vertical farms and aquaponics make food production possible even in urban centers, reducing the need for extensive land use and long-distance transportation. These systems create closed-loop environments where water is recycled, nutrients are reused, and crops are cultivated in controlled conditions, enhancing yields while conserving resources. In rural areas, energy abundance supports regenerative farming practices that restore soil health, increase biodiversity, and sustain local ecosystems. Farmers, freed from dependence on fossil fuels, harness solar and wind power to operate equipment, irrigate fields, and store produce. This integration of sustainable energy into agriculture marks a renaissance for farming, where food production and environmental stewardship are mutually reinforcing goals.

Transportation, long a significant source of emissions and resource consumption, undergoes a similar transformation in this era of energy abundance. Public transit networks powered by renewable energy become the backbone of mobility in cities and beyond, offering efficient, accessible, and sustainable alternatives to individual car use. High-speed rail networks connect distant regions, enabling swift, low-impact travel across large distances. Personal vehicles, once reliant on gasoline, are replaced by electric and hydrogen-powered alternatives that produce no emissions and are charged with renewable energy. Even aviation and shipping industries, historically dependent on fossil fuels, innovate to use sustainable fuel sources, drastically reducing their environmental impact. This reimagined transportation landscape ensures that mobility no longer strains the planet but coexists harmoniously with it, a shift that allows people to move freely without leaving a destructive footprint.

With the foundations of society now supported by renewable energy, human progress takes on a new meaning, shifting from rapid expansion to mindful enrichment. The relentless pursuit of resources that once defined progress gives way to a philosophy of balance, where growth is measured not by volume or speed but by quality and impact. This approach recognizes that humanity's greatest achievements are not those that deplete or dominate but those that contribute to the flourishing of all life. Energy abundance thus encourages a shift toward mindful development, where every project, every innovation, is guided by a commitment to enhancing life without compromising future generations.

The shift in values brought about by energy abundance permeates society, shaping cultural, economic, and personal choices alike. Individuals, empowered by access to renewable energy, adopt lifestyles that reflect a commitment to sustainability. Homes are designed with energy-efficient systems that minimize waste and conserve resources, embodying a new standard of living where sustainability is as fundamental as comfort. Communities come together to support initiatives that reduce carbon footprints, conserve water, and protect local wildlife, fostering a sense of shared responsibility. This cultural ethos extends into everyday choices, where people are encouraged to consider the environmental impact of their actions and to choose options that align with the principles of energy abundance.

Education plays a vital role in fostering this culture of sustainability and stewardship. Schools and universities, supported by renewable energy, provide students with access to resources, digital platforms, and collaborative networks that enhance learning. Curricula emphasize environmental literacy, ethics, and social responsibility, preparing students to navigate the complexities of a world where interconnectedness is paramount. Beyond academic knowledge, education becomes a journey of personal growth and social awareness, instilling in students a commitment to the well-being of the planet and the communities that depend on it. This new approach to education cultivates leaders who are not only skilled but also compassionate, individuals who are prepared to contribute to a society that values sustainability, inclusivity, and ethical progress.

As society adapts to the reality of energy abundance, the principles of governance evolve to reflect a commitment to transparency, equity, and collective well-being. In this era, governments are tasked with safeguarding the resources that sustain life, ensuring that energy is managed responsibly and distributed fairly. Policies are crafted to protect public interests, creating frameworks that prevent monopolies, support local communities, and promote environmental health. International partnerships emerge as nations recognize that the global challenges of climate change, resource management, and human rights require a unified approach. These alliances foster a spirit of cooperation, enabling countries to work together to share knowledge, protect ecosystems, and ensure that the benefits of energy abundance reach everyone, regardless of geography or wealth.

The governance model of this new era is rooted in the belief that energy, like air and water, is a universal right. This perspective demands a commitment to inclusivity, ensuring that access to energy is not limited by socioeconomic status but is available to all. To achieve this, public and private sectors work in tandem, establishing regulatory frameworks that uphold ethical standards and prevent exploitation. Community-led initiatives and public engagement become central to policy-making, creating a governance system that is accountable to the people it serves. This commitment to fair governance reflects a broader dedication to the principles of justice, unity, and environmental stewardship, principles that guide every aspect of society in the age of energy abundance.

At the intersection of technology and governance, artificial intelligence emerges as a powerful tool for managing the complexities of renewable energy systems. AI, with its ability to analyze vast datasets and optimize processes, plays a central role in ensuring that energy is used efficiently, distributed fairly, and maintained sustainably. Through predictive analytics, AI-driven systems anticipate fluctuations in supply and demand, adjust energy distribution, and minimize waste. These intelligent systems operate transparently, their algorithms designed to serve public interests and operate in alignment with ethical guidelines. By integrating AI into energy management, society is able to harness the full potential of renewable resources, creating a system that is adaptable, resilient, and responsive to the needs of people and the planet alike.

The role of AI in this era of energy abundance is not merely technical; it is ethical. As AI systems become more integral to daily life, questions about accountability, privacy, and autonomy become central to the conversation. Who has the authority to govern these systems? How can society ensure that AI is used for the public good rather than private gain? These questions underscore the need for governance models that prioritize transparency, protect individual rights, and maintain public trust. In this new age, technology is seen not as an end but as a means—a tool that, when aligned with humanity's highest values, can support a society that is fair, inclusive, and sustainable.

The transformative power of energy abundance also fosters a new relationship between humanity and the natural world, one defined by respect, reciprocity, and care. Freed from the need to exploit the Earth's resources, society shifts its focus to preservation and restoration, embracing a role as stewards rather than consumers. Conservation efforts flourish, supported by renewable energy that powers large-scale projects without depleting ecosystems. Forests, once cleared for fuel or farmland, are restored to health, creating carbon sinks that stabilize the climate. Oceans, rivers, and lakes are protected from pollution, allowing marine and freshwater ecosystems to thrive. This commitment to ecological health reflects a deep-seated recognition that the survival of humanity is bound to the well-being of the planet.

The restoration of natural ecosystems is accompanied by a cultural renaissance that celebrates the beauty, diversity, and interconnectedness of life. Artists, musicians, writers, and thinkers draw inspiration from this renewed relationship with nature, creating works that explore themes of environmental harmony, social justice, and human connection. Art becomes a medium through which society reflects on its journey, a testament to the resilience of the human spirit and the power of collective action. Public art installations, community festivals, and cultural events bring people together, creating spaces where they can celebrate the progress made and envision the path forward. In this way, the arts play a vital role in shaping society's values, reminding people of the shared responsibility they hold to protect and preserve the world for future generations.

As energy abundance reshapes every facet of life, a new understanding of human potential emerges—one that values balance, empathy, and purpose over material accumulation. The relentless pursuit of resources, once seen as the hallmark of progress, is replaced by a commitment to quality of life, community well-being, and ecological integrity. This shift in priorities redefines success, encouraging individuals to live with intention, to pursue paths that contribute to the common good, and to recognize the interconnectedness of all life. Energy abundance, in this sense, is not merely a technological achievement; it is a philosophical awakening, a call to build a world that honors the dignity of every being and the beauty of the natural world.

This new perspective on progress is reflected in the choices people make, the policies they support, and the innovations they pursue. Society embraces a lifestyle that values mindfulness and conservation, where resources are used thoughtfully and where growth is measured by its impact on the world. This ethos permeates every aspect of life, from personal habits to public policy, creating a culture that is not defined by consumption but by contribution. Communities support local food systems, renewable energy projects, and public spaces that foster connection and environmental awareness, creating a society that is resilient, inclusive, and deeply connected to the Earth.

As humanity continues to explore the possibilities of energy abundance, education becomes a cornerstone of this cultural shift. Schools, universities, and community programs emphasize global citizenship, environmental stewardship, and social responsibility, preparing students to navigate the complexities of a sustainable world. Education is no longer confined to the classroom; it is a lifelong journey that encourages individuals to grow, to question, and to engage with the world around them. This approach fosters a generation that is not only knowledgeable but also compassionate and resilient, individuals who are prepared to lead with integrity and to contribute to a society that values sustainability, inclusivity, and ethical progress.

As energy abundance transforms every corner of society, a new era of collective responsibility and unity takes shape, one that redefines humanity's relationship with progress, nature, and each other. Freed from the perpetual demands of scarcity, people, communities, and nations are encouraged to pursue pathways of growth that respect and enhance the environment rather than exploit it. This transformation invites humanity to question long-held assumptions about success, prosperity, and the purpose of progress, establishing a cultural and ethical framework that seeks not domination but harmony, not rapid accumulation but mindful development.

In this world, industries that once embodied extraction and consumption are reimagined as forces for renewal and restoration. The materials economy, long dependent on the relentless exploitation of natural resources, transitions into a model that emphasizes circularity, resilience, and longevity. Companies adopt production methods that ensure every product has a lifecycle beyond initial use, integrating design principles that prioritize recyclability, repairability, and reusability. Raw materials, once harvested with little thought to long-term impacts, are now sourced responsibly and used efficiently, minimizing waste and reducing the strain on ecosystems. In this regenerative model, each stage of production serves a purpose beyond immediate profit, contributing to an economy that values continuity and environmental health over short-term gain.

The agriculture sector, traditionally one of the most resource-intensive industries, evolves into a model of sustainable food production. Powered by renewable energy, farms integrate advanced technologies such as AI-driven crop management, water-efficient irrigation systems, and soil health monitoring to optimize growth while conserving resources. The principles of permaculture and regenerative agriculture take center stage, enabling farmers to cultivate crops in ways that restore soil nutrients, increase biodiversity, and promote ecological balance. Local food systems are strengthened as communities embrace urban farming, rooftop gardens, and community-supported agriculture, reducing dependency on long-distance food supply chains. This shift in agricultural practices not only enhances food security but also brings people closer to the source of their sustenance, fostering a culture that values the land and the people who work it.

In urban centers, energy abundance enables the creation of cities that are as environmentally supportive as they are technologically advanced. Urban design prioritizes green infrastructure, where buildings, parks, and transportation systems are seamlessly integrated to reduce energy consumption and promote sustainable living. Smart buildings, equipped with advanced energy management systems, monitor and adjust their own energy use, optimizing lighting, heating, and cooling to reduce environmental impact. Public transportation systems powered by renewable energy reduce traffic congestion, air pollution, and reliance on personal vehicles. These cities are not mere collections of buildings; they are living ecosystems that balance human needs with ecological health, embodying a vision of urban life that is both practical and beautiful, efficient and compassionate.

The transformation of cities extends into the cultural and social fabric of communities, where energy abundance fosters a new ethic of communal engagement and shared responsibility. Public spaces become gathering points for education, dialogue, and celebration, where people come together to learn about sustainable practices, celebrate cultural heritage, and discuss the future of their communities. Libraries, schools, and community centers are equipped with renewable energy resources, transforming them into hubs of empowerment where people can access information, share ideas, and participate in civic life. These spaces serve as reminders of the collective journey toward sustainability, places where individuals are encouraged to think beyond themselves and to contribute actively to the well-being of their neighbors and the environment.

Education, as a vital component of this new era, takes on a holistic approach that prepares individuals not only for professional success but for ethical leadership and responsible citizenship. Curricula emphasize the interconnectedness of ecosystems, societies, and economies, encouraging students to view their education as a means of contributing to a healthier world. Environmental science, ethics, and global studies become core subjects, teaching students about the impact of human activity on the planet and the importance of sustainable practices. This educational framework nurtures a generation that is both technically skilled and ethically grounded, individuals who are equipped to lead society toward a future that respects both human potential and ecological integrity.

As people across the globe experience the benefits of energy abundance, a shared sense of purpose and responsibility begins to shape international relations. Countries that once competed over resources and influence now recognize the value of collaboration, working together to address the universal challenges of climate change, biodiversity loss, and human rights. International partnerships and organizations play an essential role in supporting this cooperative spirit, establishing standards for sustainable development, protecting shared ecosystems, and promoting equitable access to resources. These alliances reflect a shift in global priorities, where the well-being of the planet and its inhabitants takes precedence over individual national interests. This era of global cooperation fosters peace, stability, and mutual respect, creating a foundation upon which all nations can thrive.

Governments, too, evolve in response to the demands of an energy-abundant society. Policies and regulations are crafted with a commitment to transparency, equity, and sustainability, ensuring that energy resources are managed responsibly and that their benefits are accessible to all. Public policy no longer serves only economic growth but prioritizes social well-being, environmental protection, and future generations. These policies promote clean energy access for all citizens, incentivize sustainable business practices, and protect ecosystems from degradation. This approach to governance reflects a deep commitment to the principles of energy abundance, recognizing that a truly sustainable future requires a system of laws and policies that protect the rights of people and the planet.

The technological advancements that make energy abundance possible are themselves guided by ethical frameworks that prioritize human welfare and environmental stewardship. Artificial intelligence, automation, and biotechnology are developed not only for efficiency but for the enhancement of quality of life, fairness, and ecological health. AI-driven systems, for example, are used to monitor and optimize energy use, predict environmental impacts, and support conservation efforts, ensuring that resources are used thoughtfully and that technological progress aligns with societal values. These technologies operate transparently, with clear guidelines and accountability measures that protect individual rights and maintain public trust. This ethical approach to technology reflects a recognition that true progress is measured not by speed or scale but by the extent to which it supports humanity's highest values.

The ethical frameworks that govern technology are essential in navigating the complex relationship between innovation and humanity. As society increasingly relies on AI to manage energy, transportation, healthcare, and environmental conservation, questions arise about the limits of automation, the role of human oversight, and the importance of accountability. Who decides how AI systems are developed and deployed? How can society ensure that these technologies serve the common good rather than the interests of a select few? Addressing these questions requires a commitment to transparency, public engagement, and ethical governance, creating a system where technology supports the needs and aspirations of all people.

At the heart of this transformation lies a renewed respect for the natural world, a recognition that humanity is not separate from nature but deeply interconnected with it. Energy abundance enables society to focus not on extracting resources but on preserving and restoring ecosystems, treating the environment as a partner in the journey toward a sustainable future. Forests, wetlands, oceans, and grasslands are seen not as resources to be harvested but as vital components of the Earth's ecological network, each playing a role in maintaining the balance that sustains life. Conservation efforts become central to public policy, supported by renewable energy that powers large-scale restoration projects and protects habitats from destruction.

This commitment to ecological stewardship extends to every corner of society, influencing personal choices, community actions, and cultural values. Individuals are encouraged to live in ways that respect the environment, adopting practices that reduce waste, conserve resources, and protect biodiversity. Community-led initiatives, from tree-planting programs to clean water projects, bring people together to care for their local environments, fostering a culture of environmental responsibility. Public awareness campaigns, educational programs, and media play a key role in promoting this ethic, reminding people of the beauty and fragility of the natural world and the importance of preserving it for future generations.

In this age of energy abundance, art and culture serve as powerful forces for inspiration and reflection, exploring themes of interconnectedness, sustainability, and shared responsibility. Artists, writers, and musicians create works that celebrate the Earth's beauty, confront the challenges of climate change, and envision a future defined by unity and compassion. Public art installations, performances, and cultural events bring people together, creating spaces for dialogue, celebration, and introspection. Through art, society finds a way to express its hopes, confront its fears, and celebrate its achievements, fostering a sense of shared identity and purpose.

As humanity builds a culture that values sustainability and stewardship, the very concept of progress is redefined. Success is no longer measured by the speed of development or the accumulation of wealth but by the health of ecosystems, the well-being of communities, and the extent to which society lives in harmony with the planet. This shift reflects a deep understanding of what it means to thrive, a recognition that true prosperity is found not in domination but in coexistence, not in consumption but in care. Energy abundance thus becomes a catalyst for a philosophical shift, encouraging humanity to question, to reflect, and to embrace a new vision of progress that is as ethical as it is innovative.

The journey to energy abundance is, in many ways, a journey of renewal—a return to values that honor life, respect diversity, and celebrate the interconnectedness of all things. It is a reminder that humanity's greatest achievements lie not in the conquest of nature but in the ability to live within its bounds, to adapt to its rhythms, and to protect its beauty. This vision of abundance is a call to action, a commitment to build a world where every life is valued, where every ecosystem is protected, and where every choice reflects a commitment to a sustainable future.

As the energy-abundant society continues to grow, its legacy becomes a testament to humanity's capacity for transformation. It is a story of resilience, of a civilization that chose to rise above the limitations of scarcity and to build a world defined by compassion, integrity, and unity. This legacy is not just a promise to future generations; it is a model of what is possible when society embraces the values of balance, responsibility, and shared purpose.

As the vision of energy abundance unfolds, humanity stands on the brink of a new paradigm—one in which every action, every advancement, is aligned with a shared commitment to equity, sustainability, and the well-being of all life on Earth. Freed from the constraints of scarcity, society is invited to redefine what it means to progress, to grow, and to thrive. This reimagining of human potential is not only a celebration of technological and scientific achievements but also a profound transformation of values, priorities, and interconnectedness. The age of energy abundance empowers humanity to construct a legacy that honors the planet, fosters resilience, and envisions prosperity not as a function of consumption but as a manifestation of harmony and shared purpose.

In this transformed world, energy abundance enables the flourishing of industries and practices that were once limited by the availability of resources. Renewable energy flows into every sector, supporting a reinvention of commerce, agriculture, and manufacturing that places sustainability and longevity at the forefront. This era sees the end of the throwaway culture, as businesses and consumers alike adopt models that prioritize durability, reuse, and ethical sourcing. Products are crafted with careful consideration of their environmental impact, and every item is designed to fulfill a purpose beyond its initial use. This shift in industrial values not only reduces waste but also aligns production with the natural cycles of regeneration, creating a global economy that is as restorative as it is productive.

The principles of energy abundance extend deeply into agriculture, where the very methods of food production are rethought to harmonize with nature's rhythms. Precision agriculture, supported by AI-driven insights and renewable energy, enables farmers to optimize crop yields while conserving water, maintaining soil health, and reducing chemical inputs. Energy-efficient systems allow farmers to monitor soil conditions, weather patterns, and crop growth in real time, adapting practices to protect biodiversity and prevent resource depletion. In urban centers, vertical farming and hydroponic systems produce food in controlled environments, reducing the need for arable land and minimizing transportation distances. This approach to agriculture not only meets humanity's dietary needs but also restores balance to ecosystems that sustain life, creating a food system that is resilient, localized, and environmentally conscious.

Beyond food production, energy abundance brings about a revolution in transportation and connectivity, making movement both efficient and sustainable. Public transit systems powered by clean energy become the preferred mode of transportation, reducing dependence on personal vehicles and alleviating congestion in urban areas. High-speed rail networks connect cities across vast distances, providing an alternative to air travel that is both faster and lower in emissions. Electric vehicles, once a luxury, become accessible and ubiquitous, charged by the renewable grid that powers entire cities and towns. The shift to sustainable transportation systems reduces pollution, preserves natural landscapes, and fosters a connected world where travel does not come at the expense of environmental health. In this age of energy abundance, mobility is not limited by the availability of fuel but is powered by the renewable forces that surround and sustain humanity.

The built environment, too, is transformed by the principles of energy abundance. Cities evolve into ecosystems that support both human life and biodiversity, with buildings, parks, and infrastructure designed to coexist harmoniously with the natural world. Green roofs, solar panels, and rainwater harvesting systems become standard features of urban architecture, integrating sustainable practices into the very fabric of city life. Buildings are constructed with materials that are locally sourced, renewable, and resilient, reducing the environmental impact of construction and maintenance. These urban landscapes are more than places to live and work; they are environments that support health, connection, and well-being, fostering a sense of belonging and purpose among their inhabitants. In this way, cities become symbols of humanity's commitment to living sustainably, examples of how innovation can enhance, rather than deplete, the world around us.

The transformation of cities extends into rural areas, where energy abundance enables the development of self-sufficient communities that are deeply connected to the land. Rural regions, once marginalized by limited infrastructure and access to resources, are empowered by localized energy systems that support agriculture, education, and healthcare. Microgrids, powered by solar and wind energy, allow these communities to generate their own power, reducing reliance on centralized networks and increasing resilience. Farmers and artisans thrive as local economies are revitalized, supported by the infrastructure that energy abundance provides. This decentralization of power promotes a sense of agency and pride, as people in rural areas gain the tools to contribute to a sustainable and interconnected society. In this model, rural and urban areas are not disparate; they are integral parts of a balanced system, each contributing to and benefiting from the other.

As society adapts to the realities of energy abundance, a shift in cultural values becomes evident. The relentless pursuit of growth and consumption, once synonymous with progress, gives way to a philosophy that values balance, community, and purpose. Success is no longer defined by accumulation but by the quality of contributions to the collective good, by the extent to which individuals and communities live in harmony with the world around them. People are encouraged to make choices that reflect their values, to live in ways that honor both the planet and future generations. This cultural shift creates a society where well-being is prioritized, where people are supported in pursuing paths that bring meaning and fulfillment.

This transformation in values is supported by an educational system that emphasizes environmental stewardship, ethical decision-making, and global awareness. Schools and universities cultivate a generation of leaders who are prepared to address the complexities of a sustainable society, equipping students with the skills to innovate responsibly and to act as custodians of the Earth. Curricula incorporate interdisciplinary learning, encouraging students to explore the connections between science, ethics, and social responsibility. Education becomes a lifelong endeavor, a journey of personal and social growth that empowers individuals to contribute to a world where sustainability is the norm, not the exception. This approach to education nurtures a generation that is both knowledgeable and compassionate, individuals who are dedicated to building a future that respects and preserves the planet's resources.

As the age of energy abundance continues to shape the world, governance and policy evolve to reflect the values of transparency, equity, and collective well-being. Governments are entrusted with the responsibility of ensuring that energy resources are managed fairly and ethically, creating frameworks that promote sustainable development and protect public interests. Policies support access to clean energy for all citizens, incentivize regenerative business practices, and safeguard ecosystems from exploitation. These policies are not merely regulatory; they are expressions of a commitment to a world where progress is aligned with the principles of justice, sustainability, and inclusivity. This model of governance emphasizes the importance of accountability, ensuring that decisions reflect the values of the people and the environment they serve.

The ethical frameworks that guide governance in this new era extend into the realm of technology, where artificial intelligence and automation are harnessed to support societal goals rather than private gain. AI systems are designed with ethical considerations, prioritizing applications that enhance quality of life, protect privacy, and promote transparency. These systems are used to optimize energy distribution, reduce waste, and support environmental conservation, ensuring that the benefits of technology are accessible to all. AI's role in energy management becomes one of empowerment, a tool that enables society to harness renewable resources responsibly and to maintain resilience in the face of changing conditions. This ethical approach to technology reflects a broader commitment to equity, creating a system where technological advancements serve the common good and reinforce humanity's dedication to environmental stewardship.

At the heart of this transformation lies a deep respect for the natural world, a recognition that humanity's survival is inseparably linked to the health of the planet. Conservation efforts are no longer reactive but proactive, with policies and projects aimed at preserving ecosystems and restoring damaged landscapes. Forests are reforested, wetlands are protected, and marine ecosystems are rehabilitated, creating habitats that support biodiversity and strengthen resilience to climate change. This commitment to ecological health reflects an understanding that true abundance is achieved only when the planet thrives alongside humanity. By treating nature as a partner rather than a resource, society embraces a philosophy of stewardship that values the interconnectedness of all life.

The arts, too, flourish in this age of energy abundance, serving as a powerful medium for exploring themes of sustainability, justice, and collective responsibility. Artists, writers, and musicians draw inspiration from the natural world, creating works that celebrate the beauty of Earth and the resilience of its ecosystems. Public art installations, theater performances, and community festivals bring people together, fostering a sense of unity and shared purpose. Through art, society finds a way to express its aspirations, to reflect on its challenges, and to celebrate its achievements. Culture becomes a reflection of humanity's journey, a narrative that honors the past, engages the present, and inspires the future.

As society embraces the principles of energy abundance, a new understanding of progress emerges—one that values balance, empathy, and purpose. The endless pursuit of material gain gives way to a commitment to quality of life, to the well-being of communities, and to the preservation of the environment. This shift in priorities redefines success, encouraging individuals to live intentionally, to make choices that benefit others, and to recognize the interconnectedness of all life. In this world, prosperity is not measured by the accumulation of wealth but by the health of ecosystems, the strength of community bonds, and the depth of human connection. Energy abundance, in this sense, is not only a technological achievement but a philosophical awakening, a call to build a world that reflects humanity's highest ideals.

The journey toward energy abundance is thus a journey of renewal, a reaffirmation of values that honor life, respect diversity, and celebrate the interconnectedness of all things. It is a reminder that humanity's greatest achievements lie not in the conquest of nature but in the wisdom to live within its bounds, to respect its rhythms, and to protect its beauty. This vision of abundance is a promise to future generations, a commitment to leave a world that is as rich in life, love, and learning as it is in energy and resources.

In this age of energy abundance, humanity creates a legacy that reflects its highest aspirations—a legacy of resilience, unity, and compassion. This legacy is a testament to the power of collective action, a reminder that when people come together with a shared purpose, they can create a world that honors the dignity of every life and the beauty of the natural world. Energy abundance is more than a source of power; it is a beacon that illuminates the path to a future where growth, discovery, and joy are as boundless as the energy that fuels them.

As the era of energy abundance reaches full maturity, society experiences a profound transformation that extends beyond infrastructure, economy, and technology. It is a transformation of spirit, values, and purpose—a new alignment between human civilization and the natural world that reflects the deepest aspirations of humanity. In this world, where renewable energy flows freely and sustainably, every individual, every community, and every institution embraces a commitment to harmony, compassion, and responsibility. This vision of abundance is not simply a fulfillment of material needs but a realization of a higher purpose, one that values the interconnectedness of life and strives to protect and celebrate it in every action.

The concept of progress is redefined, evolving from a narrow focus on expansion and accumulation to an enlightened understanding of prosperity. True progress, in this age, is measured by the health of ecosystems, the well-being of people, and the depth of human connection. Growth is no longer synonymous with consumption but with enrichment—of knowledge, relationships, and communities. This shift creates a society where wealth is not measured by possessions but by contributions to the collective good, by the capacity to inspire, uplift, and support one another. People are encouraged to pursue lives that bring meaning, purpose, and joy, recognizing that personal fulfillment is deeply tied to the well-being of the broader world.

Energy abundance enables humanity to cultivate not only a sustainable lifestyle but a regenerative one, where each choice and action contributes to the renewal of resources, ecosystems, and cultures. The relentless pursuit of short-term gains is replaced by a commitment to long-term stewardship, a philosophy that recognizes that the Earth's resources are gifts to be protected, cherished, and shared. This regenerative approach influences every aspect of society, from the way people consume and produce to the way they interact with each other and the environment. It is a shift from taking to giving, from ownership to stewardship, from isolation to interconnectedness.

Industries, businesses, and institutions operate on principles that prioritize sustainability, equity, and transparency. Companies understand that their success is intertwined with the health of the planet and the well-being of their communities. Corporate responsibility is not a token gesture but a foundational ethos, guiding decision-making processes that balance profitability with ethical considerations. Supply chains are designed to be local, circular, and resilient, reducing dependence on long-distance transportation and minimizing environmental impact. Products are crafted to serve purposes beyond mere functionality, designed with longevity, recyclability, and ecological integrity in mind. This approach to commerce creates an economy that values quality over quantity, resilience over rapid growth, and ethical responsibility over mere profit.

In agriculture, energy abundance fuels a revolution that makes food production more sustainable, accessible, and equitable. Farmers are equipped with technologies that allow them to monitor and manage soil health, water usage, and crop diversity, creating a food system that respects the land and supports biodiversity. Small-scale farms, community-supported agriculture, and urban farming initiatives flourish, connecting people to the source of their food and fostering a culture of gratitude for the Earth's bounty. Localized food systems reduce reliance on intensive industrial farming, creating a model of agriculture that nurtures rather than depletes. In this world, the food that nourishes humanity is cultivated with care, respect, and a recognition of the delicate balance that sustains life.

Transportation networks, powered by renewable energy, enable people to travel efficiently, cleanly, and affordably, connecting communities in ways that are respectful of both natural and human-made environments. Public transit systems reduce the need for individual car ownership, encouraging a shift toward shared mobility that reduces congestion, pollution, and urban sprawl. Electric and hydrogen-powered vehicles become the norm, producing no emissions and requiring minimal infrastructure. High-speed trains link cities across vast distances, providing alternatives to air travel that are sustainable and accessible. This transportation network fosters a connected world where people can move freely without imposing a burden on the environment, reflecting a commitment to both accessibility and ecological integrity.

Urban landscapes become living ecosystems that harmonize with the surrounding environment, blending green spaces, renewable energy, and sustainable infrastructure. Buildings are constructed to be energy-positive, generating more power than they consume, and are equipped with systems that capture rainwater, reduce waste, and provide natural habitats. Public parks, rooftop gardens, and urban forests bring nature into the heart of the city, creating spaces where people can connect with the Earth, breathe clean air, and find respite. These cities are not just centers of human activity; they are places of renewal, sanctuaries of biodiversity, and symbols of humanity's dedication to sustainable living.

In rural areas, energy abundance supports the development of self-sufficient communities that are deeply connected to the land. Localized energy systems empower these regions to generate their own power, fostering economic independence and resilience. Rural communities thrive as they become centers of sustainable agriculture, crafts, and cultural heritage, contributing to a diverse, interconnected society. By decentralizing energy production, energy abundance creates a model of self-reliant, locally focused communities that contribute to the broader ecosystem without straining it. This rural renaissance bridges the divide between urban and rural areas, fostering a society where both are valued, supported, and celebrated.

The cultural fabric of society is enriched by a collective embrace of values that prioritize mindfulness, gratitude, and responsibility. The arts, supported by the freedom and inspiration that energy abundance provides, flourish as a medium for exploring themes of justice, unity, and environmental stewardship. Artists, musicians, and writers create works that celebrate the beauty of the natural world, challenge societal norms, and inspire people to reflect on their role in protecting the planet. Public art installations, community festivals, and cultural events bring people together, creating spaces for celebration, dialogue, and collective action. Through art, society finds a way to connect, to express its hopes and fears, and to celebrate the shared journey toward a sustainable future.

Education, too, is transformed by the ethos of energy abundance. Schools, universities, and community programs foster a generation that is knowledgeable, compassionate, and ethically grounded. Curricula emphasize global citizenship, environmental literacy, and critical thinking, preparing students to navigate the complexities of a sustainable society. Education becomes a pathway not only to personal achievement but to collective progress, encouraging students to view themselves as stewards of the Earth and as members of a global community. This approach to education nurtures individuals who are dedicated to building a world that values diversity, inclusivity, and environmental integrity, creating a legacy of responsible leadership for future generations.

As the age of energy abundance unfolds, governance evolves to reflect the principles of fairness, accountability, and collective well-being. Governments are entrusted with the responsibility of managing energy resources equitably, ensuring that they are accessible to all and that their use aligns with the long-term health of the planet. Policies promote regenerative practices, incentivize ethical business operations, and protect ecosystems from degradation. Public institutions operate transparently, with input from the communities they serve, creating a governance model that is as participatory as it is effective. This model of governance embodies the belief that energy, like air and water, is a universal right, a resource to be shared and safeguarded for the benefit of all.

The role of artificial intelligence and automation in this era is one of empowerment and support. AI-driven systems optimize energy use, predict environmental impacts, and enable efficient resource management, operating transparently and with accountability. These technologies, guided by ethical frameworks, support the pursuit of societal goals rather than individual profit, aligning with humanity's commitment to equity and environmental health. AI becomes a partner in the journey toward sustainability, a tool that amplifies human potential while respecting individual rights and maintaining public trust. In this model, technology is not an end in itself but a means to achieve a higher purpose—a purpose rooted in the values of justice, compassion, and respect for all life.

The journey toward energy abundance is thus a journey of reconnection—a rediscovery of humanity's place within the web of life. Freed from the need to exploit the Earth's resources, society shifts its focus to preservation and restoration, treating the environment as a partner rather than a commodity. Conservation efforts flourish, supported by renewable energy that enables large-scale restoration projects, protects habitats, and promotes biodiversity. This era of ecological stewardship reflects an understanding that true abundance is achieved only when humanity and the planet thrive together. The commitment to living within nature's bounds creates a world where the beauty, diversity, and resilience of life are preserved and celebrated.

In this world of energy abundance, the legacy of humanity is one of unity, resilience, and compassion. It is a story of a civilization that chose to rise above the limitations of scarcity, to embrace a vision of prosperity that honors life and respects the Earth. This legacy is a promise to future generations, a commitment to leave a world that is as rich in possibility as it is in beauty. It is a testament to the power of collective action, a reminder that humanity's greatest achievements lie not in the conquest of nature but in the wisdom to live in harmony with it.

As humanity moves forward, energy abundance serves as a beacon—a guiding light that illuminates the path to a future defined by peace, inclusivity, and sustainability. This vision of abundance is a call to action, a reminder that each individual has the power to contribute to a world that honors the dignity of every life and the interconnectedness of all beings. It is a legacy of hope, resilience, and the enduring belief in the boundless potential of humanity to create a world that reflects the highest ideals of compassion, integrity, and unity.

In choosing this path, humanity commits to a future where the energy that fuels life is boundless, where resources are shared, and where progress is measured not by accumulation but by the enrichment of the human spirit and the preservation of the natural world. This is the legacy of energy abundance—a legacy that reaches beyond technology, beyond economics, and beyond politics, touching the very essence of what it means to be human. It is a vision of a world where every life is valued, every ecosystem is protected, and every step forward is a step toward a more just, harmonious, and enlightened existence.

Chapter 3: Tesla's Dream Realized—Wireless Power Transmission from Space

In a world where the limitations of physical infrastructure have long bound human potential, the concept of wireless energy transmission emerges not just as a technological solution but as a transformative vision of boundless connectivity and freedom. Rooted in the bold foresight of Nikola Tesla, this dream of energy without borders holds the power to dissolve the physical chains that restrict access, innovation, and the sharing of resources on a global scale. Tesla's ambition for wireless energy went beyond the immediate practicalities of his era; it touched on a profound reshaping of society's relationship with power itself. The realization of his ideas, now revitalized through modern advances, represents a reawakening of aspirations once thought to be confined to the fringes of science fiction.

Tesla's exploration into wireless energy transmission began with a principle as elegant as it was groundbreaking: the resonant transfer of energy through electromagnetic waves. By aligning the resonant frequencies of transmitters and receivers, he found that energy could be transmitted over distances without the need for physical connectors, wires, or direct contact. This revelation opened a door to the possibility of a world where energy could be shared freely across vast expanses, limited only by the capabilities of the transmitters and the understanding of resonance. In his mind's eye, Tesla envisioned cities aglow with power transmitted invisibly through the air, where machinery operated without cables, and homes were illuminated without the infrastructure of a conventional grid. This vision laid the groundwork for a future where humanity could transcend the constraints of physical energy networks, moving instead toward a reality of seamless, instantaneous access to power.

Tesla's initial experiments with resonant inductive coupling demonstrated the viability of this idea, revealing how electromagnetic fields could bridge gaps and transfer energy over short distances. He constructed towering resonant coils and transmitters, capable of generating powerful fields that could induce energy flow in distant receivers. His designs, although limited by the materials and technology of the time, were capable of transmitting measurable power across impressive distances, hinting at the potential for scaling this technology to a global level. However, the technology of Tesla's time imposed severe limitations. The materials available were inefficient and prone to energy loss over long distances, while the understanding of electromagnetic theory was still in its formative stages. The high voltages necessary to achieve significant transfer rates posed safety challenges, and the lack of infrastructure for regulating and directing these transmissions led to unresolved technical hurdles.

Today, advancements in material science, electromagnetism, and control systems offer solutions to many of the challenges Tesla faced, bringing his vision into the realm of practical application. Resonant inductive coupling has evolved, aided by the development of superconductors and zero-resistance materials that drastically reduce the energy losses that plagued early models. Superconducting materials, operating in ultra-cold environments, allow for near-perfect energy transfer, making them ideal for applications in wireless transmission. These materials create a pathway for energy to flow without dissipation, allowing Tesla's dream of efficient, lossless energy transmission to come closer to reality. In the vacuum of space, where these materials can operate with minimal interference, wireless energy systems find an environment conducive to large-scale, long-distance applications that were previously impossible.

Modern scientists and engineers are not only reviving Tesla's vision but are amplifying it through the concept of space-based wireless power transmission. Here, the boundaries that limited terrestrial applications of wireless energy—such as atmospheric interference, physical obstacles, and the attenuation of electromagnetic fields over long distances—can be minimized. A new infrastructure of superconducting transmitters and receivers, strategically positioned in space, offers the promise of directing energy beams from space-based energy stations to receivers on Earth. Such a system could harness the energy of the sun continuously, unhindered by the day-night cycle or weather patterns, providing a stable and abundant source of power for terrestrial use. By situating these transmission arrays in orbit or at Lagrange points, where gravitational forces reach equilibrium, these systems can maintain a constant position relative to Earth, facilitating uninterrupted energy flow with minimal adjustments.

This model of space-based wireless transmission not only extends the reach of energy beyond geographical and political boundaries but also addresses some of the most pressing energy needs of the modern world. Rural and remote regions, often isolated from conventional grids, could access power from space-based arrays, overcoming the barriers of location and infrastructure. Industrial centers, too, would benefit from uninterrupted, scalable power sources that do not deplete local resources or contribute to environmental degradation. The prospect of transmitting energy directly from space enables humanity to transcend the limitations of land-based energy production, offering a sustainable solution to the growing global demand for power.

While the vision of wireless energy transmission is compelling, it raises complex technical and ethical considerations. The challenges of aligning transmitter and receiver frequencies over vast distances require sophisticated control systems capable of real-time adjustments. Variations in distance, atmospheric conditions, and potential interferences introduce fluctuations that could disrupt the resonance necessary for efficient energy transfer. Engineers are developing adaptive systems that can detect and correct for these variations, ensuring stable energy flow regardless of environmental factors. Advances in artificial intelligence and real-time data processing allow for precise frequency adjustments that compensate for shifts in alignment, atmospheric conditions, and other variables that might disrupt transmission. This adaptability is essential for scaling wireless power transmission across vast distances, ensuring that the energy beams remain focused, directed, and stable.

Another critical consideration in the deployment of space-based wireless energy transmission is the safety of both the technology and its users. High-intensity energy beams, if misdirected, could pose risks to human health and environmental safety. Developing secure transmission protocols that prevent accidental exposure is therefore paramount. Engineers are exploring focused-beam technology that uses advanced optics and phased array systems to direct energy precisely to designated receivers, minimizing the risk of dispersion and ensuring that energy is delivered safely. These systems are designed with fail-safes that detect and redirect beams in case of misalignment or interference, safeguarding both users and ecosystems.

The deployment of space-based wireless power systems also demands careful consideration of the environmental impact, both on Earth and in space. Transmitting energy across the atmosphere requires strategic planning to avoid interference with wildlife, particularly avian and marine species that could be affected by electromagnetic fields. In space, where satellite and orbital congestion is a growing concern, the placement of wireless power arrays must consider both current and future traffic in Earth's orbit. Regulations and international agreements will play a crucial role in governing these installations, ensuring that space-based infrastructure supports sustainable and responsible development.

The potential of wireless energy transmission from space brings humanity to a turning point—a moment where the longstanding limitations of terrestrial power generation are confronted by the almost boundless opportunities presented by orbital systems. This technology, if fully realized, could redefine humanity's energy landscape, enabling societies to operate in ways previously unimagined. With the sun providing a virtually inexhaustible supply of energy, space-based power systems offer a method to capture, store, and deliver power to Earth with unparalleled reliability and minimal environmental cost. These advancements carry the promise of transforming energy from a finite, resource-bound commodity to an abundant, universally accessible force for progress and connectivity.

Space-based solar power arrays, stationed in geosynchronous orbits, can capture sunlight without the interruptions caused by weather, time of day, or seasonal variations. These arrays are engineered with photovoltaic cells optimized for high-efficiency light absorption, coupled with advanced cooling systems that maintain optimal performance even under intense solar exposure. Through concentrated arrays and specialized solar reflectors, these space-based platforms collect energy continuously, storing it in superconducting capacitors or directly converting it into high-energy beams for transmission to Earth. The persistent capture of solar energy, uninterrupted by Earth's atmospheric conditions, allows for a consistent, stable flow of power that can support a wide range of applications across industrial, residential, and rural settings.

The science of wireless energy transmission from space relies on meticulously engineered mechanisms for converting solar power into transmittable energy forms. Two primary methods dominate this field: microwave and laser-based transmission. In the microwave method, energy is converted into microwaves, which are directed toward a receiving station on Earth. Microwave beams are chosen for their relative stability over long distances and their ability to penetrate atmospheric layers with minimal energy loss. Receivers on Earth, called rectennas, capture these beams and convert them back into usable electricity, distributing it through terrestrial grids. Laser-based transmission, on the other hand, uses highly focused beams of infrared or visible light, which offer even greater energy density. However, lasers require precise alignment and clear atmospheric conditions, making them more suitable for localized transmission or as supplements to broader microwave-based systems.

Rectennas, specially designed to receive microwave transmissions, play an integral role in the success of wireless power transmission from space. These arrays of antenna elements are crafted to capture and convert the microwave beams into direct current (DC) electricity with remarkable efficiency. Placed strategically across geographical locations, rectennas offer a pathway to distribute power even to areas that are typically off the grid or isolated by challenging terrains. The design of rectennas continues to advance, with new materials and configurations allowing for higher conversion rates, lower energy loss, and improved resilience to environmental factors. As these receivers evolve, they become more adaptable, versatile, and capable of supporting the stable flow of energy from space to Earth.

While the technical possibilities are staggering, the implementation of space-based wireless energy systems introduces complex governance issues. The prospect of directing powerful energy beams toward Earth calls for international cooperation to establish standards, regulations, and protocols that ensure safety and equity. These systems, by nature, cross national boundaries, raising questions about sovereignty, resource allocation, and access rights. The development of universal guidelines that govern energy transmission and access will require a collective commitment from nations, balancing the potential benefits of wireless power with the need for responsible management.

Moreover, this technology offers a paradigm shift in the relationship between energy and power—not just in a technical sense, but in terms of social, political, and economic influence. Access to energy has historically been a point of contention, a resource often controlled by select groups or nations with the means to harness it. With wireless energy transmission, there is an opportunity to transcend the disparities that have long defined global energy distribution. A universal infrastructure for energy transmission from space could enable every nation, regardless of its geography or economy, to access a stable supply of power. However, realizing this vision requires a commitment to equity, transparency, and accountability, principles that will define the success and acceptance of space-based power systems.

The ethical dimension of space-based wireless power is equally profound. By its nature, this technology challenges the existing frameworks of resource ownership and national sovereignty. If power stations in space become centralized, operated by a few powerful entities, the risk of monopolization increases, potentially creating a new imbalance in the distribution of power. Ensuring that this infrastructure remains accessible and serves the collective interests of humanity demands a governance model that transcends traditional notions of ownership and control. Ideally, these systems would be managed as global commons, overseen by international bodies that prioritize equitable access and environmental stewardship.

Environmental considerations also extend to the potential impacts of microwave and laser beams on Earth's atmosphere, ecosystems, and human health. While researchers are developing transmission systems designed to minimize these risks, rigorous testing and continuous monitoring will be essential to prevent unintended consequences. The interaction of high-energy beams with atmospheric layers, for instance, could introduce localized heating or affect weather patterns in the vicinity of receiving stations. Scientists are actively exploring ways to mitigate these effects, developing methods for dispersing energy in ways that prevent concentrated impact on any specific area. The environmental integrity of these systems will rely on transparent research, comprehensive risk assessments, and a willingness to adapt designs as knowledge evolves.

One of the most compelling aspects of space-based wireless energy transmission is its potential to empower communities that have historically been marginalized in terms of energy access. Rural, remote, and isolated regions, which often face prohibitive costs for grid extension, could benefit directly from this technology. By placing rectennas in these underserved areas, energy from space could bypass the need for costly infrastructure, providing a sustainable solution to energy scarcity in even the most challenging locations. For countries with vast rural populations, this technology could be a lifeline, enabling access to power for development, healthcare, education, and industry. The impact of wireless energy transmission on human development could be profound, opening doors to new opportunities and improving the quality of life in regions that have long been constrained by limited energy resources.

Beyond the immediate benefits of energy access, space-based wireless transmission holds implications for global resilience and sustainability. The climate crisis and the growing frequency of extreme weather events underscore the vulnerabilities of conventional energy infrastructures. Storms, floods, wildfires, and other disasters can disrupt power grids, leading to prolonged outages and economic losses. By providing a resilient, decentralized energy source, wireless transmission from space offers a safeguard against these disruptions, ensuring continuity of power even in times of crisis. This capacity for resilience aligns with broader efforts to build infrastructure that can withstand the uncertainties of a changing climate, contributing to a future where societies are better equipped to adapt and respond to environmental challenges.

Moreover, the environmental benefits of reducing dependency on fossil fuels are amplified by the transition to space-based renewable energy. With solar arrays in space providing a significant portion of the world's power, terrestrial ecosystems are spared from the impacts of land-based energy extraction, whether through mining, drilling, or deforestation. The pressure on natural habitats decreases as renewable sources replace coal, oil, and gas, allowing ecosystems to recover and biodiversity to flourish. In this way, space-based energy systems become an instrument for conservation, aligning human progress with the preservation of the planet.

The development of wireless energy transmission from space also represents an opportunity for global unity and cooperation. This technology, by design, transcends borders, inviting nations to work together toward a common vision of sustainable energy. Collaborative research initiatives, shared technological advancements, and co-investment in space infrastructure create a framework for international partnership, fostering a culture of mutual benefit and shared progress. These alliances are not merely practical; they represent a profound shift in the way nations approach energy—moving from competition to collaboration, from control to stewardship.

As humanity approaches the realization of Tesla's vision of wireless energy, the ethical, social, and environmental dimensions of this technology become as important as the technical achievements themselves. The challenge lies not only in perfecting the mechanics of transmission but in creating a system that reflects humanity's highest ideals. It is a challenge to use this technology to uplift and empower rather than dominate, to preserve the integrity of the planet while enhancing the quality of life for all its inhabitants.

Tesla's vision, once considered too radical to be practical, now serves as a guiding light for a new era of energy innovation. In the same way that he dreamed of a world where energy flows freely and universally, space-based wireless transmission invites humanity to reimagine the possibilities of connection and power. It is a dream of a world where every community, every individual, has access to the energy needed to thrive, a world where progress is not defined by barriers but by bridges, by the open flow of resources and ideas across all divides.

As the concept of space-based wireless energy transmission continues to evolve, the underlying vision becomes one of profound interconnectedness and shared purpose. Tesla's original vision for wireless energy has expanded in scope, reshaping not only the technical ambitions of society but also its ethical, social, and ecological ideals. The path toward a world powered by energy transmitted directly from space is paved with possibilities that challenge humanity to reach beyond conventional boundaries, to reconsider how resources are shared, and to imagine a future where access to energy is no longer dictated by geography or wealth. This vision demands more than scientific rigor; it calls for a shift in mindset, an embrace of stewardship, and an unwavering commitment to principles of equity and unity.

The scientific basis of space-based energy transmission rests upon recent advancements in electromagnetism, materials science, and adaptive optics. Engineers have devised systems capable of converting captured solar energy into electromagnetic waves that can travel vast distances with minimal loss. These waves, when carefully calibrated, can penetrate Earth's atmosphere, reaching rectennas with enough precision to ensure efficiency while maintaining safety. The infrastructure required to support this technology includes not only advanced transmission arrays but also sophisticated control systems that monitor, adjust, and direct energy flows in real-time. Each transmission must be meticulously calculated, accounting for atmospheric conditions, terrestrial alignment, and environmental factors to guarantee uninterrupted energy flow without harm to ecosystems or human populations.

In practice, space-based wireless energy transmission embodies the delicate balance between technological power and environmental mindfulness. The alignment of transmission beams with receiving stations requires precision engineering that respects the natural world while achieving its ambitious aims. Scientists are developing phased array systems that allow energy beams to be steered dynamically, a feature that compensates for the subtle shifts in alignment caused by the rotation and tilt of the Earth. This real-time adaptability means that energy can be redirected as necessary, ensuring a stable supply to designated receivers without interference. Moreover, phased arrays allow for energy dispersal when needed, diffusing intensity to avoid concentrated effects in vulnerable areas. These advancements underscore the importance of harmonizing engineering with ecological awareness, building a system that is as responsive to environmental needs as it is to human demands.

The design of rectennas, which convert transmitted energy back into electricity, represents another critical area of development. Modern rectenna arrays are constructed using metamaterials—engineered substances that enhance electromagnetic responsiveness, allowing for higher energy conversion rates and greater durability in varied environmental conditions. These materials are optimized for microwave frequencies, which offer stability over long distances while maintaining efficiency. As rectenna technology advances, it becomes more resilient to weather fluctuations, adaptable to different geographies, and capable of supporting diverse energy demands. Rectennas, distributed strategically, become beacons of accessibility, empowering regions that have been historically disadvantaged in energy infrastructure.

While the technical possibilities are impressive, the broader impact of this technology transcends engineering. Space-based wireless energy transmission offers humanity a new framework for energy sovereignty—an era in which reliance on centralized, often monopolistic energy suppliers could be replaced by a distributed, equitable, and universally accessible network. However, achieving this vision requires robust governance structures that prioritize the collective good, ensuring that the benefits of wireless energy transmission are distributed fairly and transparently. The creation of international regulatory bodies, tasked with overseeing the safe and equitable use of this technology, could establish a precedent for global cooperation. Such entities would not only manage transmission protocols but also establish guidelines for energy access, preventing monopolization and encouraging inclusive development.

The principles of inclusivity and accessibility underpin the ethical framework of space-based wireless energy transmission. By making energy accessible to rural, remote, and underserved regions, this technology has the potential to bridge historical divides, creating opportunities for economic growth, education, healthcare, and industry in areas that have long been marginalized. For the first time, humanity stands on the verge of a world where every community, regardless of its physical location or economic status, can access the power needed to thrive. The prospect of extending energy access to billions of people offers a transformative vision of empowerment, one where the barriers of geography and infrastructure give way to a future defined by universal connectivity.

The potential impact on developing nations is particularly significant. For countries with limited access to natural resources or limited capital for infrastructure development, space-based energy offers a way to leapfrog the conventional path to modernization. By investing in rectenna networks and adapting existing grids to receive wireless power from space, these nations can bypass the extensive costs and environmental impacts associated with fossil fuels, dams, and large-scale land use for energy production. Instead, they can directly access sustainable, renewable power, aligning their economic growth with global environmental goals. This leapfrogging effect not only accelerates development but does so in a way that is ecologically sound, positioning these nations as active participants in the world's sustainable future.

In light of these possibilities, space-based wireless energy transmission carries profound implications for global resilience and stability. The climate crisis, marked by increasingly severe weather events, underscores the vulnerabilities of conventional energy grids. With transmission systems located in space, the impacts of hurricanes, wildfires, and floods on energy infrastructure are minimized, ensuring continuity of power even in the face of natural disasters. Wireless energy transmission offers a resilient alternative, reducing the likelihood of blackouts, disruptions, and infrastructure damage caused by extreme weather. By providing a consistent and stable energy source, these systems support society's ability to adapt and recover in the face of environmental challenges, building resilience into the very structure of human civilization.

Environmental sustainability is further supported by the reduced demand for terrestrial energy extraction. With a significant portion of power generated from space, the pressure on Earth's ecosystems is lessened, sparing forests, rivers, and oceans from the impacts of drilling, mining, and deforestation. The transition to space-based power systems offers a path toward restoring natural habitats, protecting biodiversity, and reducing pollution. This environmental reprieve enables humanity to pursue growth that does not come at the expense of the planet's resources, creating a model of development that aligns with conservation and ecological health. By lifting the burden of energy production from the Earth, space-based systems allow ecosystems to regenerate, carbon levels to stabilize, and natural landscapes to flourish.

However, the road to this vision is not without challenges, particularly in terms of governance, security, and cooperation. The potential of wireless energy to reshape global power dynamics brings with it the responsibility to establish ethical and transparent structures that prevent misuse. The concentration of power in space raises questions about security and sovereignty, as control over these systems could be exploited for political or economic leverage. To prevent this, an international regulatory framework must be established, ensuring that no single entity holds disproportionate control over space-based energy. Instead, governance should reflect the collective interests of humanity, promoting collaboration, mutual benefit, and ethical oversight.

Establishing this framework requires trust, transparency, and a commitment to cooperation. Nations, particularly those with significant space capabilities, must work together to develop protocols for the safe, equitable use of space-based energy. International bodies, potentially modeled after organizations like the United Nations, could oversee these protocols, managing access, safety standards, and dispute resolution. This governance model would act as a stabilizing force, preventing monopolies, mitigating conflict, and fostering a spirit of shared responsibility. Such a framework, if implemented effectively, could serve as a model for other areas of space activity, setting a precedent for cooperation in exploration, resource management, and environmental protection.

The ethical considerations of space-based wireless energy transmission extend beyond governance into the realm of environmental justice. As energy is transmitted across borders and into regions with historically limited access, the question of environmental impact on specific communities becomes paramount. The placement of rectennas, the alignment of energy beams, and the management of transmission frequencies all have implications for local ecosystems and populations. Researchers are exploring ways to minimize these impacts, ensuring that energy transmission systems operate within safe parameters for human and environmental health. Environmental assessments and community consultations are essential, providing a foundation for inclusive, respectful, and ecologically sound development.

For communities that have faced energy poverty, the arrival of space-based wireless power represents a promise of progress that transcends material wealth. Access to consistent, renewable energy enables these communities to invest in education, healthcare, clean water, and technological advancement, unleashing potential that has been historically constrained by lack of resources. Schools can operate with reliable lighting and connectivity, healthcare facilities can provide consistent services, and homes can be heated, cooled, and powered sustainably. The ripple effects of energy access extend far beyond convenience, laying the groundwork for human development that is both empowering and sustainable.

The cultural impact of this technology is equally profound, transforming the ways people think about energy, resource sharing, and environmental stewardship. As wireless energy transmission becomes a norm, society's relationship with power shifts from one of consumption to one of mindful use. People are encouraged to think of energy not as a commodity but as a shared resource, a force that connects and sustains communities rather than divides them. This shift in perspective fosters a culture of conservation and respect, where energy is used thoughtfully and responsibly, and where individuals are aware of their impact on the planet. In this way, space-based wireless energy transmission is more than a technological breakthrough; it is a catalyst for cultural change, fostering values of unity, sustainability, and collective responsibility.

The potential of space-based wireless energy transmission, therefore, lies not only in its ability to provide clean power but in its capacity to inspire a new vision of progress. This vision is one in which energy abundance serves as a foundation for peace, equity, and environmental harmony. It invites humanity to transcend the limitations of scarcity, to imagine a world where resources are shared freely and where the power to innovate and thrive is accessible to all. As nations and communities embrace this technology, they join in a global project that promises not only technical advancement but a reimagining of what it means to be interconnected, responsible, and human.

The realization of space-based wireless energy transmission marks a profound leap in humanity's journey toward a world defined by connectivity, sustainability, and equity. Each step toward this vision not only brings new technological achievements but also prompts society to reconsider the ethical foundations of progress. As Tesla's dream of wireless power begins to take shape in the form of space-based systems, the potential for transformative change touches all areas of human life, from individual well-being and environmental health to global stability and cooperation. The development and deployment of these systems are guided by a commitment to creating a future that transcends traditional notions of energy as a scarce commodity, forging instead a reality where energy flows freely and inclusively across boundaries, empowering every corner of society.

One of the most profound aspects of space-based energy transmission is its potential to mitigate the energy inequities that have long divided the world. Access to energy has historically been a determinant of economic opportunity, with communities and nations lacking energy infrastructure often unable to participate fully in the global economy. By making clean, renewable energy universally available, space-based transmission systems have the power to change this dynamic, empowering regions that have long been isolated by geography or limited by lack of resources. Rural communities, remote islands, and economically disadvantaged areas could gain access to energy without the costly and invasive construction of traditional infrastructure, leveling the playing field and offering new opportunities for development, innovation, and growth.

In this context, the arrival of wireless energy from space holds promise as a tool for environmental justice, addressing the unequal impacts of energy scarcity and environmental degradation. Communities that have historically borne the brunt of pollution and environmental harm could benefit from the shift away from land-based fossil fuels and extractive practices. The reduction in demand for coal, oil, and natural gas would relieve the environmental pressures on these communities, allowing for the restoration of natural habitats, improvement of air quality, and rejuvenation of local ecosystems. The health benefits of this transition are equally significant, with cleaner air, water, and land contributing to improved quality of life, reduced disease, and enhanced resilience for communities worldwide.

Beyond addressing energy inequity, space-based wireless transmission offers a pathway to climate stability by dramatically reducing the global dependence on fossil fuels. Solar arrays positioned in space capture energy continuously, unimpeded by atmospheric conditions, cloud cover, or the Earth's rotation. This constant energy supply enables a degree of reliability and scalability that is challenging to achieve with terrestrial renewables alone. By supplementing ground-based solar, wind, and geothermal systems, space-based power can serve as a stable backbone for the global energy grid, providing a consistent supply that can offset fluctuations from other sources. This reduction in fossil fuel use translates directly to lower carbon emissions, supporting international efforts to combat climate change and reduce humanity's ecological footprint.

The environmental benefits of space-based energy transmission extend beyond reduced emissions, encompassing broader ecological impacts. The transition from extractive industries to renewable energy sources allows ecosystems that have been disrupted by mining, drilling, and deforestation to recover. Forests once cleared for coal or biofuel production can regenerate, oceans can heal from the effects of offshore drilling, and arable land previously used for energy crops can return to food production or biodiversity conservation. This environmental restoration is not only a boon for wildlife but also strengthens resilience against the impacts of climate change, as healthy ecosystems play a vital role in stabilizing the climate, regulating water cycles, and preserving biodiversity.

In this new era, energy abundance challenges humanity to reimagine its relationship with resources, prompting a cultural shift toward mindful, sustainable living. When energy flows freely and cleanly from space, the fear of scarcity is replaced by a sense of responsibility and conservation. Individuals and communities are encouraged to think beyond immediate consumption, considering the broader implications of their energy use and the potential impact on future generations. This cultural shift toward conscious energy use fosters a society where conservation is not only a matter of policy but a personal and collective value, embedded in everyday choices and actions. By creating a world where energy is abundant and accessible, space-based systems foster a culture of respect for the Earth's finite resources and the intricate balance that sustains life.

The promise of energy abundance also extends to the fields of science and technology, where access to stable, renewable power unlocks new realms of possibility. Research facilities, hospitals, universities, and laboratories that require substantial, uninterrupted energy to operate can function without the limitations imposed by local power grids. This access enables advancements in medical technology, genetics, environmental science, and countless other fields, accelerating the pace of discovery and innovation. Scientists working on climate adaptation, genetic research, and space exploration, for instance, can conduct experiments with the confidence that they are supported by a reliable energy infrastructure. The synergy between sustainable energy and scientific inquiry fosters an environment where progress is bound only by imagination, where the next breakthrough in health, sustainability, or technology is always within reach.

In education, the impact of space-based energy transmission extends beyond the walls of research institutions, reaching schools, universities, and digital learning platforms worldwide. With energy barriers removed, educational institutions in remote and underserved regions gain access to the resources and connectivity that support modern education. Rural schools can power computers, internet access, and digital tools that enable interactive learning and access to global knowledge. Online education platforms, powered by renewable energy, become more accessible, allowing students to learn from anywhere, at any time, creating a more inclusive educational landscape. This leveling of the educational playing field empowers future generations to engage in lifelong learning, equipped with the tools to tackle the challenges of the future.

The integration of space-based wireless energy systems also brings humanity closer to a circular economy—a model in which resources are used, recycled, and reintegrated with minimal waste. By reducing reliance on finite resources, space-based energy supports an economy that prioritizes resilience and sustainability over rapid consumption. Industries are encouraged to adopt practices that minimize environmental impact, designing products that are durable, repairable, and recyclable. The shift to a circular economy aligns with the goals of space-based power systems, creating a cycle of production that respects the limits of the planet while maximizing the utility and lifespan of every resource. This economic model supports both ecological health and economic stability, offering a vision of prosperity that endures beyond the limits of finite resources.

The potential for global unity inspired by space-based wireless energy transmission is equally profound. By transcending the limitations of geography and physical infrastructure, this technology invites humanity to approach energy not as a commodity to be fought over but as a shared resource that belongs to all. Nations that once competed for access to fossil fuels now have the opportunity to work together to develop, maintain, and share energy from space-based systems. This shift in perspective fosters cooperation, reducing tensions over energy security and encouraging international collaboration. Joint initiatives to develop and manage wireless energy infrastructure create opportunities for cultural exchange, mutual learning, and strengthened diplomatic relations, building a foundation for peace rooted in shared interests.

The governance of space-based energy systems requires unprecedented cooperation, transparency, and ethical stewardship. As nations and organizations work together to design and deploy these systems, they face the challenge of balancing sovereignty with collective benefit, ensuring that no single entity dominates the distribution or control of energy from space. International agreements, modeled after existing treaties on space and environmental protection, could establish frameworks for equitable access, safety protocols, and environmental safeguards. These agreements would outline responsibilities, establish standards, and create mechanisms for accountability, ensuring that space-based energy remains a force for good in the world.

To address the technical and ethical challenges of space-based energy, international regulatory bodies could be empowered to oversee research, set guidelines, and mediate disputes. Such bodies would ensure that the development of space-based systems aligns with the values of equity, sustainability, and transparency. By holding stakeholders accountable to these principles, regulatory frameworks would protect the public interest, prevent monopolization, and promote the responsible use of space resources. These efforts not only safeguard the integrity of space-based energy but also create a model for cooperation that could guide future endeavors in space exploration, resource management, and environmental protection.

In the broader context of humanity's journey, space-based wireless energy transmission represents not only a technical achievement but a philosophical milestone. It embodies a vision of a world where progress is defined by unity, where the pursuit of knowledge and innovation serves the common good, and where the natural and human worlds exist in balance. The journey toward this vision is one of interconnectedness, a recognition that the well-being of humanity and the health of the planet are inseparably linked. Space-based energy transmission is more than a means of powering human civilization; it is a testament to the potential of humanity to transcend old limitations, to rise to new challenges, and to create a world that reflects the values of compassion, resilience, and shared responsibility.

The legacy of this technology is one that invites future generations to think beyond the constraints of scarcity and conflict, to imagine a world where energy is not a source of division but a catalyst for unity. It is a legacy that challenges humanity to approach every resource, every invention, and every opportunity with a sense of responsibility and reverence for the planet that sustains us. By embracing the possibilities of space-based energy, humanity sets a course toward a future where technological advancement is balanced with ethical commitment, where the pursuit of progress respects both the Earth and all its inhabitants.

In this world of energy abundance, societies are freed from the limitations that have historically constrained them, empowered to pursue their highest aspirations and to forge a future defined by hope, equity, and environmental stewardship. Space-based wireless energy transmission, born from Tesla's dream and realized through generations of scientific ingenuity, offers a model of development that values not only the power to innovate but the wisdom to use that power responsibly. It invites humanity to build a world where every choice, every advancement, and every action reflects the enduring commitment to a sustainable, just, and connected global community.

As space-based wireless energy transmission moves from vision to reality, the full implications of this transformative technology unfold across every sphere of life, reshaping not only how humanity accesses power but also how it conceives of community, sustainability, and progress. This leap toward energy abundance redefines the possibilities of human existence, casting aside the long-standing constraints of scarcity and shifting focus to a future of boundless opportunity, shared resources, and collective responsibility. In this new paradigm, energy becomes more than a utility; it becomes a symbol of unity, resilience, and the potential for a harmonious relationship between humanity and the Earth.

The technological foundation of space-based energy transmission rests on the sophisticated infrastructure that collects, converts, and delivers solar power from orbit to the Earth's surface. Space-based arrays, positioned strategically to capture sunlight continuously, provide a stable source of energy that is unaffected by the day-night cycle or seasonal variations. These arrays operate as immense solar farms in the vacuum of space, optimized for efficiency and durability. The collected solar energy is converted into microwave or laser beams, which are directed toward Earth-based receivers with unparalleled precision, ensuring minimal energy loss and safe transmission. This technological feat is achieved through adaptive optics and phased arrays, which allow for real-time adjustments to maintain alignment and energy flow, even as the Earth rotates and atmospheric conditions fluctuate.

The rectenna fields on Earth that receive and convert this transmitted energy represent another marvel of engineering, designed to maximize efficiency while integrating seamlessly into the local environment. These rectennas, constructed with high-resilience materials capable of enduring various climates, convert microwaves or laser beams into electricity with remarkable efficacy. The resulting power is then fed into the grid or distributed locally, bringing renewable energy directly to communities, industries, and households without the need for extensive physical infrastructure. As these fields become operational, they offer a visual testament to the realization of Tesla's dream, a landscape where energy flows invisibly yet constantly, illuminating homes, powering factories, and supporting digital connectivity across the globe.

The advent of space-based energy transmission redefines energy as a shared, equitable resource rather than a contested, finite commodity. This new approach invites humanity to break free from cycles of extraction, conflict, and environmental degradation, instead embracing a model where energy is harnessed from an inexhaustible source and distributed without regard for borders or wealth. By providing a sustainable solution to energy access, space-based systems have the potential to end energy poverty, democratizing access to power in ways that allow for true global development. As the cost of energy becomes a non-issue, nations previously hindered by high energy costs can channel their resources into healthcare, education, infrastructure, and innovation, fostering a new era of equitable growth.

In this new energy landscape, humanity's relationship with the environment is reimagined. As space-based power reduces the need for terrestrial energy production, the demand for fossil fuels, hydropower, and even some forms of biomass decreases, allowing ecosystems to recover and flourish. Forests, once cleared for energy-related land use, are preserved or restored, becoming carbon sinks that support biodiversity and combat climate change. Rivers that would have been dammed are allowed to flow freely, maintaining the natural habitats essential to aquatic life. The reduced pressure on Earth's resources allows for a shift from exploitation to regeneration, where human activity complements and enhances the planet's ecological balance rather than disrupting it. This shift reflects a broader cultural movement toward sustainability, where progress is aligned with preservation, and growth is synonymous with care for the environment.

The environmental benefits of space-based energy transmission also extend to the mitigation of climate change. By decreasing reliance on fossil fuels, space-based power significantly reduces carbon emissions, contributing to global efforts to stabilize the climate. With a reliable, continuous energy source that does not produce greenhouse gases, nations can transition to a low-carbon economy, achieving their climate targets and supporting a sustainable future. The potential for space-based energy systems to serve as a primary or supplementary source of power during peak demands or in areas with limited renewable resources enables a flexible approach to climate resilience, ensuring that the energy grid can adapt to fluctuations without compromising the health of the planet.

This technology also brings a profound shift in economic structures and labor markets. The development, maintenance, and management of space-based energy systems create new opportunities in fields such as aerospace engineering, renewable energy, AI-driven control systems, and environmental science. Jobs in these fields offer pathways for economic empowerment, especially in regions that have traditionally depended on resource extraction. As fossil fuel industries gradually phase out, workers are supported in transitioning to roles in renewable energy, space technology, and sustainable development, with governments and educational institutions offering training programs to facilitate this shift. This transition is a testament to the adaptability and resilience of human labor markets, demonstrating that economic growth can be compatible with environmental stewardship.

The economic transformation inspired by space-based energy is not limited to developed nations. Developing countries, which often face barriers to industrialization due to high energy costs and limited infrastructure, find in this technology an equalizer. With reliable access to affordable, clean energy, these nations can pursue economic growth without the environmental costs associated with fossil fuels. Industries in sectors such as manufacturing, technology, and agriculture can flourish, creating employment opportunities and supporting local economies. By freeing these nations from the constraints of costly energy imports, space-based power allows them to chart their own development paths, guided by principles of sustainability and self-sufficiency.

The education sector, too, benefits immensely from this transformation. With energy access no longer a barrier, schools in rural and underserved regions gain the ability to operate without interruption, powering digital learning tools, internet connectivity, and modern educational facilities. The digital divide narrows as students across the globe gain access to resources that were once limited to more developed areas. Online education platforms, enriched by stable, renewable power, become accessible to all, fostering an environment of lifelong learning and global collaboration. This inclusive approach to education empowers students in even the most remote areas, ensuring that no child is deprived of the knowledge and opportunities they need to thrive in a connected world.

The infrastructure supporting space-based energy transmission also strengthens societal resilience against crises and natural disasters. Traditional energy grids are vulnerable to disruptions from extreme weather events, earthquakes, and other catastrophes. By diversifying energy sources to include space-based power, societies create an energy infrastructure that is robust and adaptable. When terrestrial grids are compromised, wireless energy from space can provide emergency support, powering hospitals, communication networks, and essential services. This resilience not only safeguards lives but also supports recovery efforts, allowing affected regions to rebuild more quickly and effectively. In this way, space-based energy transmission becomes a cornerstone of disaster preparedness, offering a lifeline when conventional systems falter.

In fostering resilience and promoting equity, space-based wireless energy transmission also serves as a bridge for international cooperation. The development, management, and sharing of space-based energy resources invite nations to work together toward common goals, fostering a spirit of unity and shared responsibility. Collaborative research initiatives, joint investments in infrastructure, and cross-border energy sharing arrangements create bonds that transcend political divides. By prioritizing mutual benefit over competition, these partnerships offer a model for addressing other global challenges, such as climate change, biodiversity loss, and resource scarcity. This technology demonstrates that the most pressing issues of our time are best addressed not in isolation, but through collective action that respects the interconnectedness of all nations and ecosystems.

As humanity steps into this new era, the philosophical implications of space-based wireless energy transmission come to the forefront. The realization of this technology symbolizes a profound shift in the understanding of power—not just electrical power but the very notion of influence, control, and progress. In this new paradigm, power is no longer concentrated in the hands of a few but is distributed as a shared resource, a common good that supports collective well-being. The democratization of energy challenges the traditional hierarchies and power structures that have long defined global relationships, opening the door to a world where equity, transparency, and responsibility guide decision-making processes.

The cultural impact of this shift is equally significant, inspiring a generation to think beyond their immediate needs and to consider the legacy they wish to leave. With the specter of scarcity removed, society is free to cultivate values of mindfulness, compassion, and stewardship. Energy, once a source of anxiety and limitation, becomes a reminder of humanity's capacity for innovation, resilience, and ethical growth. Communities are encouraged to adopt practices that honor both the planet and future generations, fostering a culture where progress is measured not by material accumulation but by the health and harmony of the world. This shift in values represents a collective awakening, a movement toward a future where humanity lives in balance with the Earth, mindful of its role as a steward rather than a conqueror.

The legacy of space-based wireless energy transmission is thus a legacy of interconnectedness, a reminder that humanity's greatest achievements lie in its ability to work together, to overcome challenges, and to reach for a future that benefits all. As the technology matures, it serves as a beacon for what is possible when society chooses cooperation over competition, equity over dominance, and sustainability over exploitation. This legacy invites future generations to continue the work of building a world that honors the interconnectedness of all life, a world where every choice reflects a commitment to peace, resilience, and shared prosperity.

In this vision of the future, energy abundance is not an end but a beginning—a foundation upon which humanity can build a world that is fair, just, and enduring. It is a promise to future generations, a commitment to leaving a world that is as rich in beauty, opportunity, and hope as it is in resources. This legacy of space-based wireless energy transmission is more than a technological advancement; it is a statement of faith in humanity's potential to create a world that reflects the best of what it means to be human. It is a call to action, a challenge to approach every resource, every technology, and every choice with a sense of responsibility, compassion, and reverence for the Earth and its inhabitants.

The culmination of space-based wireless energy transmission represents not just the completion of a technological feat but the dawn of a new paradigm, where the power to drive human progress is matched by a commitment to ethical responsibility, environmental harmony, and shared prosperity. As humanity harnesses the sun's energy from beyond the confines of Earth's atmosphere, it gains more than a renewable resource; it inherits a legacy defined by unity, resilience, and a deep respect for the planet. The journey to this point reflects humanity's boundless curiosity and resolve, a testament to the enduring spirit of innovation tempered by a recognition of its profound impacts on life and ecosystems alike.

In this world of energy abundance, individuals, communities, and nations find themselves bound not by the limitations of scarcity but by the shared potential for a sustainable future. Space-based energy transmission reshapes civilization's approach to growth, liberating societies from the ecological and economic constraints of terrestrial energy sources. Free from reliance on finite fuels and environmentally intrusive infrastructure, economies around the globe experience a revitalization that prioritizes resilience, inclusivity, and regeneration. This shift allows each society to pursue prosperity on its own terms, free from the need to compromise environmental integrity for development.

At the community level, the effects of this technology are transformative. For the first time in history, rural and underserved regions gain access to reliable, clean energy that transcends geographical isolation. Local industries, agriculture, education, and healthcare all flourish as energy barriers dissolve, enabling these communities to fully participate in the global economy and to improve their quality of life. By addressing energy poverty, space-based power systems create a pathway for millions to rise out of hardship, offering opportunities that extend beyond basic needs to include education, health, and the potential for self-determined growth. For children in remote villages, hospitals in isolated areas, and small businesses previously constrained by limited access to power, this technology opens doors that once seemed out of reach, transforming lives and communities in ways that reverberate across generations.

Industries are equally transformed, with entire sectors reoriented around principles of sustainability and resource stewardship. The ready availability of clean, renewable energy from space empowers industries to innovate without compromising environmental health. Manufacturing processes are designed with energy efficiency in mind, minimizing waste and maximizing durability. This sustainable model of production shifts away from extractive practices, promoting instead a circular economy where resources are reused, recycled, and returned to the ecosystem. This industrial transformation not only reduces pollution and resource depletion but also fosters economic stability, as industries build systems that prioritize long-term resilience over short-term gain.

The environmental impact of space-based energy transmission extends far beyond the reduction of carbon emissions. As humanity reduces its reliance on fossil fuels, ecosystems across the planet begin to recover, and natural landscapes are preserved. Forests, freed from the pressures of deforestation for energy production, are allowed to flourish as vibrant habitats that sequester carbon, support biodiversity, and contribute to climate stability. Wetlands, rivers, and marine environments experience reduced pollution, allowing species that have been threatened or displaced to return and thrive. This reprieve from environmental strain reflects a shift in humanity's role from dominator of nature to steward of it, embracing a model of growth that coexists with the natural world rather than exploiting it.

At a planetary scale, the shift to space-based energy underscores the potential for technology to serve as a force for conservation rather than consumption. By sourcing energy from outside the Earth's biosphere, humanity can alleviate the relentless demand for resources that has historically defined industrial progress. This new approach enables society to embrace regenerative practices that restore and protect ecosystems, addressing the environmental damage of the past while fostering a more sustainable future. The reduction in land, water, and atmospheric pollution fosters a world where ecological restoration becomes as fundamental as economic development, positioning environmental health as a pillar of human progress.

This environmental rejuvenation is accompanied by a cultural transformation that reshapes humanity's relationship with both energy and each other. Energy, once a source of conflict and inequality, becomes a unifying force, symbolizing humanity's capacity to cooperate, innovate, and share. Individuals come to view energy as a common good, a shared resource that binds them to a larger community, both human and ecological. This cultural shift encourages a mindset of conservation and mindfulness, where the use of resources is guided by a commitment to equity, sustainability, and respect for future generations. As societies recognize the interconnectedness of their actions, they cultivate values that celebrate harmony, stewardship, and the ethical use of technology.

The influence of space-based energy extends into governance, reshaping the frameworks through which nations interact and collaborate. As energy ceases to be a point of competition and scarcity, governments find new grounds for cooperation, working together to manage, distribute, and protect the shared resource of space-based power. The regulation and oversight of these systems call for unprecedented levels of transparency, accountability, and collective responsibility. International alliances form to oversee the deployment and management of space-based energy systems, creating a governance model that prioritizes fairness, environmental protection, and the equitable distribution of benefits. This global framework for energy governance not only ensures that space-based power serves the public good but also sets a precedent for managing other shared resources, from water and air to space and biodiversity.

The governance of space-based energy also inspires innovation in the realm of policy and diplomacy. Nations, recognizing the mutual benefits of shared energy resources, engage in treaties and agreements that protect the infrastructure of space-based systems and regulate access to power. These treaties become foundational documents that reflect a collective commitment to peace, cooperation, and the ethical management of technology. In this new world, where energy flows freely across borders, international relations are redefined by trust and shared goals rather than competition and control. The management of space-based power becomes a testament to humanity's ability to transcend divisions, fostering a model of diplomacy that values interconnectedness and mutual benefit over division and dominance.

At an individual level, this era of energy abundance empowers people to live in alignment with their values, embracing lifestyles that prioritize sustainability, ethical consumption, and global citizenship. The accessibility of clean energy enables individuals to make choices that reduce their environmental impact, from powering their homes with renewable resources to supporting industries that uphold principles of ethical production. This lifestyle shift is not driven by sacrifice but by a sense of purpose, as individuals recognize that their choices contribute to a larger movement toward a sustainable and equitable world. Energy abundance offers the freedom to pursue fulfillment, creativity, and connection, allowing people to focus on experiences, relationships, and contributions to their communities rather than the pursuit of material accumulation.

The educational sector, empowered by space-based energy, nurtures a generation that understands the interconnectedness of life, technology, and the environment. Schools, universities, and digital platforms are equipped with the resources needed to foster environmental literacy, critical thinking, and global awareness. Students are taught to see themselves as stewards of both the Earth and the shared resources that sustain it, instilling in them a sense of responsibility that transcends borders. This approach to education prepares young people not only for careers but for lives of purpose and contribution, empowering them to address the complex challenges of a world where technology and ethics intersect. By fostering a culture of learning that values sustainability, equity, and innovation, the educational system builds a foundation for a future defined by informed, compassionate, and resilient leaders.

In the arts, literature, and cultural expression, the influence of space-based energy is celebrated as a symbol of humanity's capacity for harmony and creative aspiration. Artists, writers, and thinkers draw inspiration from the possibilities of a world powered by the sun, exploring themes of connection, responsibility, and unity. Public art installations, performances, and festivals celebrate the vision of a society where energy is abundant, sustainable, and inclusive, creating spaces where people can reflect on their role in this shared journey. The arts become a medium through which society expresses its hopes, confronts its fears, and imagines a future where technology serves not as a barrier but as a bridge to deeper connection and understanding.

The legacy of space-based wireless energy transmission is thus a legacy of possibility—a testament to humanity's capacity to dream beyond the constraints of the present and to act in ways that honor both the Earth and future generations. This legacy challenges humanity to use its most powerful technologies to uplift, to protect, and to unite, building a world where the pursuit of progress is guided by empathy, responsibility, and foresight. The journey to energy abundance is a reminder that humanity's greatest achievements are those that respect the delicate balance of life, honor the interconnectedness of all beings, and create pathways for every individual to thrive.

In choosing this path, humanity commits to a future where the energy that fuels civilization is as boundless as the sun itself, where resources are shared freely, and where progress is measured not by the accumulation of power but by the enrichment of life in all its forms. This vision is not merely a technological aspiration but a philosophical one—a statement of belief in humanity's potential to create a world that reflects the highest ideals of compassion, integrity, and unity. Space-based wireless energy transmission is a beacon, illuminating a path to a future where every life is valued, every ecosystem is preserved, and every action is a step toward a more just, harmonious, and enlightened existence.

As humanity steps into this new era, it leaves behind the limitations of scarcity, conflict, and environmental degradation, embracing instead a legacy of abundance, peace, and shared purpose. This legacy is a gift to future generations, a promise to leave a world that is as rich in opportunity, diversity, and beauty as it is in resources. It is a vision of a world where technology and nature coexist in harmony, where energy is a bridge rather than a barrier, and where the collective wisdom of humanity guides every step forward. In this world, progress is defined not by power over others but by the power to uplift, to heal, and to create a future that honors the Earth and all its inhabitants.

Chapter 4: A World Beyond Energy Scarcity—The Societal and Environmental Implications

In a world replete with complex systems and intricate networks, the role of energy goes beyond mere fuel; it becomes the heartbeat of modern civilization, an invisible yet omnipresent force shaping every facet of human existence. The exploration into understanding energy's more profound roles in society, beyond its physical and technological implications, opens a discourse on how societies prioritize resources, how industries grow or wither, and ultimately, how civilization itself either thrives or falters. As humanity grapples with the realities of energy abundance, the question arises: How does a civilization recalibrate its ethos and infrastructure to align with a world no longer bound by scarcity?

The vision of a society that no longer struggles for resources, a vision rooted in the ideals of energy abundance, calls for a paradigm shift in humanity's collective priorities and ambitions. Such a society would need to redefine the purpose of growth, progress, and technological advancement, not as pursuits of dominance over nature or other nations but as efforts in harmony, aimed at enhancing quality of life, preserving ecological balance, and ensuring sustainable well-being for all. The path to this new paradigm is one of transformation, a journey away from the reactive, consumption-driven models that have historically guided human development and toward a vision that places well-being, balance, and renewal at its core.

The adoption of energy abundance prompts a fundamental reevaluation of industry and infrastructure. In traditional models of development, industries have often operated on the assumption of limited resources, fostering a relentless drive for extraction, production, and expansion. However, a society where energy is abundant no longer needs to adhere to such models.

Instead, industries can operate within frameworks that prioritize efficiency over quantity, resilience over rapid growth, and sustainability over unchecked consumption. This shift is not merely economic; it is an ethical and environmental transformation, one that repositions industry as a partner to nature rather than a competitor. In an era defined by limitless clean energy, the concept of progress is measured not by volume or velocity but by the quality of its impact and the depth of its contributions to society and the environment.

In this transformed industrial landscape, manufacturing, technology, and resource management are redesigned to reflect the principles of a circular economy. Products are crafted to be durable, repairable, and recyclable, with each component designed for a lifecycle that aligns with the regenerative capacity of the environment. Factories are powered by renewable sources, their operations guided by efficiency and precision rather than sheer output. Resources, once extracted and discarded in cycles of waste, are now reintegrated into new products or returned to the Earth in ways that replenish rather than deplete. This circular approach reflects a profound shift in values, a commitment to using human ingenuity to support and sustain life rather than exploiting it.

Energy abundance enables a level of technological innovation that was previously unimaginable, empowering humanity to rethink the very design of cities, communities, and transportation systems. Freed from the constraints of fossil fuels, cities become centers of environmental harmony, with buildings that generate more energy than they consume, public spaces that support biodiversity, and transportation networks that operate without pollution. In this new vision of urban life, architecture and infrastructure are inseparable from nature, each element designed to minimize ecological impact and maximize community well-being. Buildings are equipped with energy-efficient systems that monitor and optimize their own consumption, reducing waste and supporting local energy grids. Public transit systems, powered by clean energy, allow for seamless mobility, connecting people and places in ways that are efficient, accessible, and environmentally friendly.

These cities, built on the foundation of energy abundance, become examples of resilience and adaptability, places where technology and nature coexist in a mutually supportive relationship. Green spaces, community gardens, and natural water management systems are woven into the fabric of urban design, creating environments that are as functional as they are beautiful. The value of these spaces extends beyond their aesthetic appeal; they represent a commitment to ecological integrity, a recognition that the health of a city is inseparably linked to the health of its ecosystems. This holistic approach to urban design creates cities that are not only livable but regenerative, places where human activity enhances rather than diminishes the natural world.

In rural areas, energy abundance transforms agricultural practices, allowing farmers to produce food in ways that are sustainable, efficient, and respectful of the land. Freed from the need for intensive chemical inputs, agriculture becomes a model of regenerative practices, where soil health, water conservation, and biodiversity are prioritized. Energy-efficient systems such as precision irrigation, soil monitoring, and renewable-powered equipment allow farmers to optimize their operations, producing higher yields with less impact on the environment. The emphasis shifts from maximizing output to cultivating balance, creating agricultural systems that work with nature rather than against it. This approach not only supports food security but also fosters economic resilience, allowing rural communities to thrive without compromising the resources that sustain them.

As energy abundance reshapes agriculture, it also enhances the resilience of food systems, creating networks that are less vulnerable to climate change and other external shocks. Localized energy production enables communities to sustain themselves independently of centralized grids, reducing the risk of disruption from extreme weather events, political instability, or supply chain interruptions. Farmers, empowered by renewable energy, can adapt to changing conditions, using data-driven insights and adaptive technology to respond to shifts in weather, soil conditions, and crop demands. This adaptability is crucial in a world where environmental uncertainty is the new norm, allowing agriculture to continue supporting human life in ways that are flexible, sustainable, and ecologically responsible.

The implications of energy abundance extend far beyond the realms of industry and agriculture, reaching into the very heart of human culture and identity. As societies embrace a model of progress that values quality over quantity, the relentless drive for accumulation gives way to a philosophy of balance and intentionality. People are encouraged to pursue paths that enhance not only their own well-being but also the well-being of their communities and the planet. This cultural shift reflects a new understanding of what it means to thrive, a recognition that true prosperity lies not in the endless pursuit of more but in the mindful stewardship of what is available. In this world of energy abundance, individuals and communities are empowered to live with purpose, to make choices that contribute to a sustainable future, and to prioritize the things that truly matter: relationships, creativity, and the joy of living in harmony with the world.

In education, this transformation is particularly evident, as schools, universities, and community programs adopt curricula that emphasize environmental literacy, ethical reasoning, and global citizenship. Students are encouraged to explore the connections between science, society, and the environment, developing a holistic understanding of their role as stewards of the planet. This educational approach fosters a generation of leaders who are not only skilled but also compassionate, individuals who are prepared to use their knowledge to contribute to a sustainable and equitable world. By emphasizing the values of sustainability, responsibility, and empathy, education becomes a force for positive change, preparing students not only for careers but for lives of purpose and impact.

This redefined approach to education extends beyond formal institutions, permeating communities and shaping the way people think about their place in the world. Public awareness campaigns, community workshops, and media programs reinforce the principles of energy abundance, encouraging individuals to adopt sustainable practices in their daily lives. Energy literacy becomes a fundamental skill, empowering people to understand and manage their own energy consumption, to make choices that reflect their values, and to participate actively in the transition to a sustainable future. This collective commitment to education and awareness fosters a culture of mindfulness, where people are conscious of the impact of their actions on the world and motivated to contribute to the common good.

The shift in values brought about by energy abundance also redefines the nature of progress, transforming the ways in which societies measure success. In this new paradigm, growth is not synonymous with expansion but with enrichment; progress is not measured by economic output but by quality of life, environmental health, and social well-being. This redefinition of progress reflects a deeper understanding of humanity's role in the world, a recognition that the well-being of individuals is inseparably linked to the well-being of communities, ecosystems, and future generations. By placing sustainability, resilience, and equity at the center of development, societies create a model of progress that is both aspirational and achievable, a vision of a world where human ingenuity and ecological integrity coexist in harmony.

The redefinition of progress in a world powered by energy abundance brings with it a profound shift in societal priorities, guiding humanity toward a future where technological advancement serves a purpose greater than individual gain. This purpose is defined by a commitment to balance, justice, and the preservation of life in all its forms. As humanity steps into this new era, it finds that the achievements worth pursuing are those that enhance collective well-being, protect the natural world, and support a sustainable existence. Freed from the constraints of resource scarcity, societies no longer need to sacrifice ecological health for economic growth, nor are they forced to choose between innovation and responsibility. Instead, a new equilibrium is forged, one that honors the interconnectedness of all life and embraces the ethical use of knowledge and power.

In this transformed society, technology becomes an instrument of harmony rather than a tool for domination. The potential of artificial intelligence, robotics, and automation is unlocked not for the purpose of reducing labor costs or maximizing productivity at any cost but for enhancing quality of life, improving accessibility, and supporting sustainable practices. AI-driven systems are developed with ethical considerations at their core, designed to serve the public good, to enhance transparency, and to operate in ways that are aligned with human values. These technologies are not designed to replace human intelligence but to complement it, supporting people in tasks that are challenging, hazardous, or repetitive, while preserving opportunities for creativity, connection, and meaningful work.

The application of AI in energy management exemplifies this ethos of technological stewardship. With the complexities of energy grids, storage systems, and distribution networks, the potential for AI to optimize and balance these systems is immense. AI algorithms monitor real-time energy flows, predict demand patterns, and adjust distribution to minimize waste and ensure that energy is directed where it is needed most. This intelligent management of energy resources allows communities to operate with efficiency and precision, reducing the strain on infrastructure and supporting the stability of renewable sources. The ethical design of these systems ensures that they remain accountable to the people they serve, with safeguards that protect privacy, transparency, and public trust.

Beyond energy management, AI also plays a critical role in environmental conservation, supporting efforts to monitor, protect, and restore ecosystems. Using data from satellite imagery, environmental sensors, and biodiversity records, AI systems can track changes in habitat, detect illegal activities such as deforestation or poaching, and assess the health of critical ecosystems. This capacity for large-scale environmental monitoring empowers conservationists, scientists, and policymakers to respond swiftly to emerging threats, to allocate resources efficiently, and to make informed decisions that support ecological resilience. The integration of AI in environmental protection is a testament to the potential of technology to serve as a guardian of life, a partner in humanity's commitment to preserve the planet for future generations.

The ethical frameworks that govern AI and automation in this age of energy abundance reflect a broader societal commitment to equity and justice. In a world where resources are no longer scarce, humanity finds itself in a position to address longstanding inequalities, to dismantle the systems that have perpetuated disparities, and to build a future that offers opportunity for all. Energy, once a tool of control and competition, becomes a shared resource, a foundation upon which a just society is built. This shift in perspective fosters policies that prioritize inclusivity, supporting communities that have been historically marginalized or underserved. Through investments in renewable infrastructure, education, and social programs, societies strive to create environments where every individual has the resources and support needed to thrive.

Access to energy, in this equitable world, is recognized as a fundamental human right rather than a privilege afforded only to those with wealth or political power. Policies are enacted to ensure that every household, regardless of location or income, has access to reliable, affordable, and sustainable power. This commitment to universal energy access brings light, warmth, and connectivity to communities that were once isolated, creating opportunities for economic growth, education, and social engagement. The impact of this universal access is profound, as people in remote or impoverished areas gain the means to improve their quality of life, to engage in commerce, to pursue education, and to connect with the world around them.

The transformation of energy into a universal right also fosters a new model of governance, one that prioritizes transparency, accountability, and collective well-being. Governments are tasked with overseeing energy resources responsibly, creating policies that balance economic growth with environmental protection and social equity. Public agencies work in tandem with private companies and community organizations to establish frameworks that support sustainable development, protect ecosystems, and ensure that the benefits of energy abundance are accessible to all. This model of governance operates with a commitment to the public interest, recognizing that the management of energy resources is not merely a matter of economics but a moral responsibility to future generations.

In this reimagined governance structure, communities play an active role in the decision-making processes that shape their energy landscapes. Local councils, community groups, and public forums provide spaces where citizens can voice their opinions, participate in policy discussions, and contribute to the shaping of energy systems that reflect their needs and values. This approach to governance fosters a sense of agency and accountability, as people are empowered to participate actively in the stewardship of their own resources. By prioritizing community engagement, societies create systems that are not only efficient but also resilient, adaptable, and aligned with the principles of equity and environmental justice.

The transparency that defines governance in this era of energy abundance is further supported by advances in data sharing, public monitoring, and open-source platforms. Energy consumption, production, and environmental impact data are made accessible to the public, allowing individuals, researchers, and organizations to hold decision-makers accountable. This open flow of information promotes trust between governments and the people they serve, creating a culture of transparency that extends beyond energy to all aspects of public life. By making information freely available, societies encourage a culture of inquiry, analysis, and accountability, empowering citizens to make informed choices and to engage in meaningful dialogue about the future they wish to create.

As societies embrace energy abundance, the principles of global cooperation and mutual aid emerge as guiding ideals. The recognition that energy is a shared resource fosters partnerships between nations, where expertise, technology, and resources are exchanged for the benefit of all. In this new spirit of cooperation, wealthier nations support energy infrastructure projects in developing regions, recognizing that the stability and well-being of one nation are inseparably linked to the well-being of others. This approach to international relations reflects a commitment to peace, unity, and the understanding that the challenges of the modern world—climate change, resource scarcity, and economic inequality—can only be addressed through collaboration.

International organizations and treaties play a critical role in coordinating these efforts, establishing guidelines for the ethical use, distribution, and management of space-based energy resources. By creating standards that govern everything from environmental impact to access rights, these organizations help to ensure that energy systems operate within ethical and ecological boundaries. This global governance framework creates a foundation for trust, a platform upon which nations can build cooperative initiatives that promote both environmental protection and social equity. In this world, where energy is abundant and shared, borders fade into the background, replaced by a vision of a world where every individual has the resources they need to thrive.

This vision of energy abundance as a catalyst for global unity also supports a renewed focus on cultural exchange, knowledge sharing, and mutual respect. As nations work together to harness and distribute space-based energy, they find opportunities to learn from each other, to celebrate diverse perspectives, and to build a sense of shared identity. Collaborative research initiatives, joint educational programs, and cultural exchange initiatives foster a spirit of understanding and solidarity, creating connections that transcend political, economic, and geographical divides. This exchange of ideas and values enriches all participants, promoting a culture of empathy, curiosity, and a willingness to engage with the complex, interconnected world.

The global perspective encouraged by energy abundance also inspires a deeper sense of responsibility toward the environment. In a world where resources are no longer perceived as finite, humanity is called to adopt a philosophy of stewardship, one that respects the Earth's natural limits and strives to preserve its beauty and biodiversity. This commitment to environmental protection becomes a defining characteristic of progress, an essential measure of any nation's success. Societies invest in conservation efforts, protect endangered habitats, and restore ecosystems, recognizing that the health of the planet is essential to the health of all its inhabitants. By treating nature as a partner rather than a resource, humanity cultivates a model of growth that is as regenerative as it is prosperous, a vision of progress that honors the interconnectedness of all life.

In the realm of education, the values of stewardship, equity, and sustainability are woven into the fabric of curricula, fostering a generation of learners who are both knowledgeable and compassionate. Schools and universities provide students with the tools they need to navigate an increasingly complex world, equipping them with skills in critical thinking, ethical reasoning, and environmental science. Beyond technical knowledge, education instills in students a sense of purpose, a commitment to using their talents and abilities to contribute to a world that values justice, peace, and the protection of the planet. This approach to education prepares students not only for careers but for lives of service, empowering them to become leaders who will carry the torch of sustainability and equity into the future.

The emergence of energy abundance as a transformative force in society elevates the pursuit of knowledge and understanding, empowering humanity to redefine its relationship with both technology and the natural world. In this new paradigm, progress is not measured by the relentless pursuit of resources but by the depth of collective insight, the resilience of communities, and the sustainability of ecosystems. Knowledge, once a means to dominate and exploit, becomes a pathway to harmony, a bridge that connects people across borders, generations, and ideologies in the shared endeavor of safeguarding the Earth and nurturing human potential. The principles of energy abundance invite humanity to embrace a life guided by purpose and foresight, one where every innovation, every advancement, and every choice is informed by a commitment to ethical growth and ecological responsibility.

The application of energy abundance in science and research further fuels humanity's capacity to understand and interact with the universe in meaningful ways. With access to limitless renewable energy, research facilities, laboratories, and exploration initiatives can operate without interruption, free from the constraints that once limited their reach. In this world, scientific inquiry is no longer confined to privileged institutions but becomes accessible to scholars, innovators, and learners around the globe. By democratizing access to energy, societies unlock the potential of human creativity and ingenuity, fostering a culture of discovery that transcends socioeconomic boundaries. Individuals from diverse backgrounds are empowered to contribute to scientific advancements, to question, to innovate, and to explore, enriching the collective understanding of the world and advancing knowledge in ways that benefit all.

Space exploration, too, takes on a renewed sense of purpose in the context of energy abundance. With the ability to generate, store, and transmit energy from orbital arrays, humanity's reach extends further into the cosmos, enabling missions to distant planets, moons, and asteroids that were once beyond reach. Spacecraft powered by sustainable energy explore the solar system, capturing data that broadens humanity's understanding of the universe and its origins. These missions are no longer driven solely by the pursuit of resources or national prestige but are grounded in a collective aspiration to expand human knowledge, to understand the cosmos, and to search for signs of life beyond Earth. This era of exploration fosters a sense of unity among nations, a shared quest to answer the mysteries of the universe and to carry the legacy of humanity's curiosity and resilience into the stars.

The vast potential of energy abundance also transforms healthcare, making it possible to provide essential services to communities that have long faced barriers to access. Medical facilities, powered by renewable energy, operate continuously, ensuring that healthcare is available to all who need it, regardless of location or economic status. Remote villages and isolated communities are no longer cut off from life-saving treatments, as mobile health clinics, telemedicine, and medical drones provide a link to trained professionals and advanced diagnostics. The availability of clean, reliable power enables hospitals and clinics to maintain critical services, power life-saving equipment, and support the infrastructure necessary for effective healthcare. This transformation of the healthcare landscape underscores the ethical imperative of energy abundance, where technology serves to uphold human dignity and to extend the promise of health and well-being to every individual.

In this world of energy abundance, the role of artificial intelligence in healthcare further exemplifies the potential of technology to support humanity's highest values. AI-driven diagnostics, predictive analytics, and personalized treatment plans offer doctors and patients insights that improve outcomes and reduce the strain on healthcare systems. Algorithms trained on vast datasets provide early warning systems for disease outbreaks, monitor patients' vital signs in real-time, and recommend tailored treatment plans that align with each individual's unique medical profile. The integration of AI into healthcare is guided by principles of ethics, privacy, and respect for human autonomy, ensuring that technology enhances, rather than replaces, the patient-doctor relationship. This approach to healthcare reflects a model where every advancement is evaluated not only by its efficacy but by its ability to support human dignity and well-being.

Education, too, experiences a renaissance in this era, as energy abundance enables institutions to operate in ways that are more accessible, inclusive, and innovative. Schools in rural and underserved regions gain the ability to offer digital resources, internet access, and modern educational tools, closing the digital divide and ensuring that every child has access to quality education. Learning becomes a lifelong pursuit, with online platforms, virtual classrooms, and community programs providing individuals of all ages with opportunities to expand their knowledge, to develop new skills, and to engage with a world of ideas. This model of education fosters a culture of curiosity, where learning is not confined to a particular stage of life but is celebrated as a continuous journey of growth, reflection, and self-improvement.

The curricula in these transformed educational systems emphasize global citizenship, environmental stewardship, and social responsibility. Students are taught to think critically about the impact of their actions, to understand the complexities of ecological systems, and to engage with the challenges of sustainability and social justice. This approach to education equips individuals with the skills they need to contribute positively to society, fostering a generation of leaders who are not only knowledgeable but also compassionate and committed to creating a world that values balance, justice, and resilience. Education in this new paradigm is not merely a preparation for the workforce but a preparation for life, instilling in students a sense of purpose and a commitment to ethical living.

In the arts, the influence of energy abundance sparks a resurgence of creativity and expression, as artists explore themes of interconnectedness, resilience, and the human relationship with nature. Freed from the constraints of resource scarcity, artists have the means to experiment, to innovate, and to challenge societal norms through their work. Public art installations, performances, and cultural festivals become celebrations of life, community, and sustainability, creating spaces where people can reflect on their shared journey and envision the world they wish to build. The arts become a vital force in shaping societal values, reminding people of the beauty and fragility of the world, and inspiring them to live in ways that honor and protect it.

The cultural impact of energy abundance extends into everyday life, influencing how people think about consumption, relationships, and personal fulfillment. In a world where resources are plentiful and sustainable, the relentless pursuit of material wealth gives way to a philosophy of minimalism, mindfulness, and gratitude. People are encouraged to value experiences over possessions, to cultivate relationships over status, and to live in ways that reflect their respect for the planet and each other. This shift in values fosters a culture where well-being is prioritized, where people find fulfillment not in accumulation but in connection, creativity, and the pursuit of meaning. The principles of energy abundance thus become woven into the very fabric of society, shaping how people work, play, and engage with the world.

This cultural transformation also redefines the nature of economic success, guiding businesses and industries to adopt models that prioritize sustainability, ethics, and social impact. Companies are no longer measured solely by profit margins but by their contributions to the well-being of communities and the environment. Corporate responsibility takes on new meaning, as businesses embrace transparent practices, fair labor standards, and environmental stewardship. The integration of renewable energy into operations allows companies to reduce their carbon footprint, minimize waste, and contribute to a circular economy that respects the planet's finite resources. In this world, success is defined not by short-term gains but by the ability to create lasting value for society, the environment, and future generations.

The principles of energy abundance also inspire new forms of economic cooperation, where industries, governments, and communities work together to address shared challenges. Collaborative networks of businesses and organizations share resources, knowledge, and technology, creating a cooperative economy that benefits all participants. This model of cooperation reduces inefficiencies, supports innovation, and promotes resilience, allowing industries to adapt to changing conditions and to contribute to a stable, sustainable economy. By prioritizing collaboration over competition, societies create an economic environment that values collective well-being over individual gain, fostering a sense of shared responsibility and mutual support.

In governance, the influence of energy abundance creates a new model of leadership, where policymakers are guided by principles of transparency, accountability, and ethical stewardship. Governments, freed from the pressures of resource scarcity, are able to make decisions that reflect the long-term interests of the planet and its people. Policies are designed to protect ecosystems, to promote social equity, and to support sustainable development, creating a framework that ensures the responsible use of resources for generations to come. This model of governance reflects a commitment to public service, a recognition that leadership is not a privilege but a responsibility to serve the common good.

The transparency that defines this model of governance is supported by advances in technology, which enable citizens to participate actively in the political process. Public platforms, open data initiatives, and community forums provide spaces where individuals can engage with policymakers, contribute their perspectives, and hold leaders accountable. This approach to governance fosters a culture of trust and engagement, where people are empowered to shape the decisions that affect their lives and the future of the planet. By prioritizing inclusivity and participation, societies create a model of governance that is not only effective but also reflective of the values of equity, respect, and collective responsibility.

This transformed society, empowered by energy abundance, embraces a vision of progress that values life in all its forms, that recognizes the interconnectedness of all beings, and that seeks to build a world that is just, sustainable, and filled with possibility.

The journey toward this vision is not without challenges, but it is guided by a collective commitment to ethical growth, to respect for the planet, and to the pursuit of a future where every individual has the opportunity to thrive. As humanity steps into this new era, it does so with a renewed sense of purpose, a recognition that true progress is measured not by what is gained but by what is shared, protected, and sustained.

The realization of energy abundance brings humanity into an era where the very fabric of existence is woven with the threads of sustainability, shared purpose, and ethical growth. With the limitless flow of renewable energy, societies gain the opportunity to move beyond mere survival, reaching toward a model of progress that is mindful, regenerative, and inclusive. This transformation is as profound as it is far-reaching, influencing not only the external structures of industry, infrastructure, and economy but also the internal values, ambitions, and dreams of individuals and communities. In this world, the relentless drive for consumption gives way to a deeper exploration of meaning, purpose, and connection, redefining what it means to thrive.

The shift from scarcity to abundance alters the core principles of social and economic systems, establishing a framework where cooperation, resilience, and equity become the primary indicators of success. Industries, once driven by the need to maximize output at the lowest cost, now operate within paradigms that value ecological health, social well-being, and ethical practices. Resource extraction, mass production, and wasteful disposal—hallmarks of a bygone era—are replaced by closed-loop systems, sustainable materials, and regenerative processes. The shift to circular economies becomes not just a technological achievement but a philosophical commitment, one that recognizes the inherent worth of resources and the importance of preserving them for future generations.

In this new economy, companies are encouraged to innovate in ways that align with societal values, creating products that are durable, recyclable, and designed to enhance rather than detract from the environment. Industries that once operated with minimal regard for their ecological impact now find themselves accountable to both communities and the planet, with transparency and responsibility forming the backbone of corporate identity. Through this alignment with ethical standards, businesses gain not only public trust but also a renewed sense of purpose, transforming from entities driven solely by profit to institutions that contribute to the well-being and resilience of the society that supports them. This transformation is further fueled by the realization that true success is not measured by financial gain but by the positive impact on people and the planet.

As these principles of sustainability take root, the concept of work itself evolves, guided by a new understanding of value and fulfillment. The relentless pursuit of efficiency, which once led to automation and job displacement, is tempered by a commitment to human dignity, creativity, and purposeful engagement. Jobs that were once seen as obsolete or replaceable are reimagined in ways that prioritize human skills, ingenuity, and compassion. In fields such as education, healthcare, environmental restoration, and the arts, roles are valued not for their contribution to economic output alone but for their capacity to enrich lives, foster connection, and build a more just society. Work becomes a space for personal growth, a means of contributing to the collective good, and an avenue for realizing one's potential within a supportive community.

The impact of energy abundance on the environment is equally transformative, creating opportunities to restore, protect, and celebrate the natural world. Freed from the dependence on finite resources, societies adopt practices that respect the limits of ecosystems, operating within the regenerative capacity of the Earth. Deforestation, mining, and other extractive activities are minimized, allowing forests, rivers, and other natural landscapes to recover and flourish. This restoration of nature is not only a benefit to the planet but also a gift to future generations, a legacy of care that speaks to humanity's responsibility as stewards of the Earth. By protecting biodiversity, preserving habitats, and reducing pollution, societies embrace a model of coexistence that honors the intrinsic value of every species, recognizing that the well-being of all life forms is interconnected.

The preservation of natural spaces extends beyond the practical benefits of ecosystem services; it reflects a cultural shift toward reverence, awe, and gratitude for the beauty of the natural world. In cities, green spaces, parks, and urban forests are designed to bring nature into daily life, creating environments where people can experience the tranquility and inspiration that nature provides. These spaces serve as reminders of humanity's relationship with the Earth, fostering a sense of connection that transcends the boundaries of urban and rural, human and non-human. This integration of nature into urban design transforms cities into living ecosystems, places where human activity harmonizes with natural processes and where people are encouraged to live in ways that reflect respect for the planet.

The arts play a crucial role in this cultural shift, providing a space where society can reflect on its journey toward sustainability and express its hopes for the future. Artists, musicians, writers, and performers draw inspiration from the themes of interconnectedness, resilience, and renewal, creating works that celebrate life in all its forms. Public art installations, community performances, and festivals become avenues for collective reflection, bringing people together to explore the values and ideals that define this new era. Through their work, artists inspire a sense of wonder, curiosity, and responsibility, reminding society of the beauty of the world and the importance of preserving it for generations to come.

This cultural renaissance, driven by the values of energy abundance, extends into the realm of education, where curricula are reimagined to emphasize environmental literacy, social responsibility, and ethical reasoning. Students are taught not only the technical skills they need to succeed but also the values and principles that guide responsible citizenship. They learn to see themselves as stewards of the Earth, to think critically about the impact of their choices, and to understand the interconnectedness of global systems. This holistic approach to education prepares young people to contribute meaningfully to a world that values sustainability, equity, and compassion, equipping them with the knowledge, skills, and mindset to navigate the complexities of the future.

The integration of environmental and ethical considerations into education is supported by advances in technology, which make it possible to create learning environments that are inclusive, engaging, and adaptable. Online platforms, virtual classrooms, and interactive tools allow students from diverse backgrounds to access quality education, regardless of their location or economic circumstances. This democratization of knowledge empowers individuals to pursue lifelong learning, to explore new fields, and to develop a sense of agency over their own growth and development. By making education accessible and relevant, societies create a foundation for continuous improvement, a culture of curiosity and exploration that supports both personal and collective advancement.

The model of governance that emerges in this era of energy abundance is one that reflects the values of transparency, accountability, and collective well-being. Leaders are no longer driven by short-term gains or narrow interests; instead, they are committed to creating policies that support sustainable development, social equity, and environmental stewardship. Government agencies work in collaboration with private sector partners, community organizations, and international bodies to create policies that align with the principles of sustainability, ensuring that resources are managed responsibly and that the benefits of energy abundance are shared equitably. This collaborative approach to governance fosters a culture of trust, as people see that their leaders are working in their best interests and that the systems in place are designed to promote the common good.

Transparency in governance is further reinforced by advances in digital communication, which enable citizens to participate actively in the decision-making process. Public forums, open data initiatives, and community consultations provide avenues for individuals to engage with policymakers, to contribute their perspectives, and to hold leaders accountable. This inclusive approach to governance fosters a sense of ownership and responsibility, as people feel empowered to shape the future of their communities and to work together toward shared goals. By creating a government that is open, accessible, and responsive, societies build a model of leadership that values inclusivity, accountability, and the pursuit of a just and sustainable world.

In this world, the concept of prosperity is redefined, shifting away from material wealth toward a holistic understanding of well-being, community, and connection. Prosperity is measured not by the accumulation of goods but by the quality of life, the health of ecosystems, and the strength of relationships. People are encouraged to find fulfillment in meaningful work, creative expression, and the pursuit of knowledge, rather than in the relentless pursuit of wealth. This redefinition of prosperity creates a society that values experiences over possessions, relationships over status, and purpose over profit. In this way, energy abundance supports a culture that prioritizes balance, well-being, and the collective good, creating a world where people live in harmony with each other and the planet.

The impact of this cultural shift is profound, as individuals find new sources of satisfaction and meaning in their lives. People are encouraged to slow down, to appreciate the beauty of the world around them, and to cultivate a sense of gratitude for the opportunities and resources they have. This slower, more mindful approach to life fosters a culture of contentment and resilience, where people are able to cope with challenges, to find joy in simple pleasures, and to live in ways that reflect their values and priorities. The principles of energy abundance thus become woven into the fabric of society, shaping not only how people live but also how they relate to each other, to the planet, and to the future.

As societies embrace this model of progress, they find themselves empowered to build a legacy that reflects the highest ideals of humanity. This legacy is one of stewardship, compassion, and wisdom, a testament to humanity's capacity to create a world that honors the dignity of every life and the beauty of the natural world. The journey toward energy abundance is more than a technological achievement; it is a moral and philosophical awakening, a commitment to living in ways that respect the Earth and uplift each other. This legacy invites future generations to continue the work of building a world that reflects the best of what it means to be human, a world where progress is measured not by what is taken but by what is given back.

The journey toward a world sustained by energy abundance unfolds as a profound reimagining of what human progress can achieve when rooted in the values of stewardship, equity, and interconnectedness. As societies adapt to this new paradigm, the measure of success becomes not the accumulation of wealth or power but the quality of life experienced by individuals, the resilience of communities, and the integrity of ecosystems. The principles guiding this transformation—conservation, ethical responsibility, and a commitment to future generations—establish a framework within which humanity can flourish without compromising the health of the planet. In this new era, human ingenuity is directed not toward the exploitation of resources but toward fostering a world where prosperity and sustainability exist in harmony.

This shift toward ethical growth is reflected across every aspect of society, where the pursuit of abundance is tempered by a sense of responsibility and purpose. Industries that were once focused on maximizing profit and productivity are reoriented toward practices that are regenerative, resilient, and respectful of natural limits. In the energy sector, this commitment to sustainability drives innovation in areas like decentralized energy production, localized storage systems, and smart grids, which allow communities to generate, store, and manage their own energy. This localized approach reduces reliance on centralized systems, fosters energy independence, and ensures that every community has the power it needs to thrive. Energy, once a centralized and controlled commodity, becomes a universal resource, accessible to all and shaped by the needs of local ecosystems and societies.

As energy independence grows, so does the capacity for communities to shape their own futures. With access to reliable, renewable power, communities gain the means to develop industries that align with their unique cultural and environmental landscapes. Local economies, freed from the constraints of energy scarcity, flourish with the creation of small-scale industries, artisanal trades, and ecotourism ventures that reflect the values of sustainability and cultural preservation. These industries are not designed for mass production or rapid expansion; instead, they are rooted in the principle of quality over quantity, fostering a model of economic growth that values community well-being, cultural heritage, and environmental stewardship. This shift creates economies that are as diverse as they are resilient, driven not by the demands of a global market but by the aspirations of local communities.

In agriculture, the impact of energy abundance enables a transformation from conventional, intensive farming methods to regenerative practices that restore soil health, conserve water, and support biodiversity. The use of renewable energy in agriculture allows farmers to adopt technologies such as precision irrigation, drone-assisted crop monitoring, and vertical farming, which maximize efficiency while minimizing environmental impact. By embracing regenerative agriculture, communities become custodians of the land, creating food systems that sustain both people and the planet. This approach ensures that food production is aligned with natural cycles, reducing the need for chemical inputs, preserving vital ecosystems, and fostering resilience against climate-related disruptions. As communities reconnect with the land, they build food systems that are not only sustainable but also culturally and ecologically enriching.

This deepened connection to the land is mirrored in the way communities approach water resources. With the power of renewable energy, water management systems are designed to be efficient, adaptable, and responsive to local needs. Communities harness solar or wind-powered desalination, wastewater treatment, and rainwater harvesting systems to provide clean water sustainably, reducing dependence on distant sources and preserving natural watersheds. This localized approach to water management supports the health of aquatic ecosystems, mitigates the risk of drought, and ensures that every community has access to the water it needs to thrive. As water becomes a shared resource rather than a contested one, communities build systems that reflect the principles of conservation, equity, and responsibility, nurturing a future where access to clean water is a universal right.

The ethical frameworks guiding these transformations are echoed in the education system, where students are encouraged to explore the connections between science, society, and the environment. Curricula are designed to cultivate environmental literacy, critical thinking, and ethical reasoning, preparing students not only to excel academically but also to contribute meaningfully to a sustainable world. Teachers emphasize the importance of understanding ecological systems, the impact of human activity on the environment, and the values of responsible citizenship. By fostering a culture of curiosity and compassion, education becomes a pathway to a future where individuals are equipped to address the complex challenges of a world defined by interdependence. This educational approach instills in students a sense of purpose and empowers them to become advocates for a sustainable and equitable future.

In this new paradigm, education is not limited to formal institutions but is woven into the fabric of community life, with public awareness campaigns, community workshops, and hands-on learning experiences reinforcing the principles of sustainability. Knowledge-sharing initiatives enable communities to exchange ideas, skills, and practices that support environmental stewardship and social well-being. This inclusive approach to education ensures that every individual, regardless of age or background, has the opportunity to learn, grow, and contribute to the common good. By democratizing access to knowledge, societies create a foundation for collective advancement, fostering a culture where every member is an active participant in the journey toward a sustainable future.

The principles of transparency and accountability that define this era are further supported by technological advancements that enable open data sharing, public monitoring, and community engagement. Citizens have access to real-time information on energy production, environmental impact, and resource management, allowing them to make informed decisions and to hold policymakers accountable. This open flow of information strengthens the bond of trust between governments and the people they serve, fostering a sense of collective ownership and responsibility. By creating systems that are transparent and participatory, societies empower individuals to shape the policies that affect their lives, ensuring that decisions reflect the values and priorities of the community.

In governance, these principles translate into a model of leadership that values inclusivity, accountability, and the long-term well-being of society. Leaders are guided by a vision of progress that is aligned with the principles of sustainability and equity, creating policies that support regenerative practices, protect ecosystems, and promote social justice. Government agencies collaborate with private organizations, research institutions, and local communities to establish frameworks that balance economic growth with environmental protection and social equity. This approach to governance reflects a commitment to serving the common good, recognizing that the health of a society is inseparably linked to the health of the planet.

The role of international cooperation is also transformed in this era of energy abundance, as nations recognize the shared responsibility to protect global resources and to promote peace, equity, and sustainability. Collaborative agreements establish guidelines for the ethical use, distribution, and management of resources, creating a framework for shared stewardship that transcends borders. This commitment to international cooperation reflects a recognition that the challenges of the modern world—climate change, resource scarcity, and economic inequality—can only be addressed through unity and mutual aid. By working together, nations build a foundation for global resilience, a future where prosperity is a shared endeavor, and where every community has the resources it needs to thrive.

The global perspective encouraged by energy abundance also inspires a renewed appreciation for cultural diversity, knowledge sharing, and mutual respect. Collaborative research initiatives, joint educational programs, and cultural exchanges create connections that transcend political, economic, and geographical divides. These partnerships foster a spirit of understanding and solidarity, enriching all participants and promoting a culture of empathy, curiosity, and inclusivity. This exchange of ideas and values encourages societies to celebrate their unique identities while recognizing their shared responsibilities, creating a world where diversity is a source of strength and resilience.

At the individual level, this cultural transformation inspires people to live with intention, gratitude, and a sense of responsibility toward future generations. Freed from the constraints of scarcity, individuals are encouraged to pursue lives that reflect their values, to cultivate relationships, and to contribute to the well-being of their communities. People are empowered to make choices that prioritize sustainability, to support businesses that uphold ethical standards, and to engage in practices that respect the planet's resources. This shift toward mindful living fosters a culture where people find fulfillment not in the relentless pursuit of material wealth but in the joy of connection, creativity, and contribution.

In this era of energy abundance, humanity builds a legacy that reflects its highest ideals, a testament to the potential for ethical growth and compassionate stewardship. This legacy is one of hope, resilience, and shared purpose, a commitment to creating a world that honors the dignity of every life and the beauty of the natural world. As humanity moves forward, the principles of energy abundance become a guiding light, illuminating a path toward a future where progress is measured not by the accumulation of power but by the enrichment of life, the preservation of ecosystems, and the unity of communities.

This vision of abundance is a call to action, a reminder that each individual has the power to contribute to a world that values balance, sustainability, and harmony. It is a legacy of resilience and the enduring belief in humanity's ability to create a world that reflects the best of what it means to be human. This journey toward energy abundance is more than a technological advancement; it is a philosophical commitment, a choice to live in ways that honor both the Earth and future generations. In this world of energy abundance, societies find the freedom to pursue not only prosperity but also peace, equity, and the promise of a sustainable future.

As humanity advances along the path of energy abundance, the profound transformations brought by this shift become an enduring testament to human resilience, ethical innovation, and unity. In a world where energy flows freely and sustainably, societies are empowered to cultivate a holistic approach to progress that honors every dimension of life. This era redefines prosperity, valuing the health of ecosystems, the vitality of communities, and the dignity of individuals. It is a journey that emphasizes coexistence over consumption, stewardship over dominance, and collective growth over individual accumulation. The very fabric of civilization is rewoven with threads of compassion, sustainability, and shared purpose, creating a legacy that transcends borders and generations.

This renewed approach to prosperity encourages individuals and communities to reimagine the role of technology and resources in their lives. Energy, once a limited and contested commodity, becomes an enabler of possibility, empowering every sector of society to innovate responsibly. In healthcare, the power of renewable energy allows hospitals and clinics to operate consistently, bringing life-saving care to remote regions and urban centers alike. Medical devices run uninterrupted, telemedicine connects patients to specialists from afar, and diagnostics reach new levels of precision and accessibility. In this world, healthcare is not a privilege but a universally accessible right, supported by an infrastructure that reflects humanity's commitment to well-being and dignity for all.

The values guiding healthcare also resonate in the field of environmental stewardship, where energy abundance enables a more profound commitment to conservation and restoration. Freed from the need to extract resources from fragile ecosystems, societies dedicate efforts toward repairing the environmental harm of the past and safeguarding biodiversity. Forests, wetlands, oceans, and grasslands are protected and nurtured, becoming sanctuaries of natural beauty and biodiversity. Conservation initiatives gain unprecedented support, empowering communities to protect endangered species, restore degraded landscapes, and create new reserves that support both human and ecological health. This era of energy abundance fosters a world where humanity and nature coexist harmoniously, where each species and ecosystem is recognized for its intrinsic value and contribution to the global balance of life.

The impact of energy abundance on cities is equally transformative, as urban areas become spaces where innovation and nature converge in harmony. Buildings are designed to generate more energy than they consume, seamlessly integrating solar panels, green roofs, and sustainable materials into their architecture. Public transportation systems, powered by renewable energy, allow people to move through the city without contributing to pollution or traffic congestion. Parks, green corridors, and urban farms provide residents with access to nature, offering spaces for recreation, connection, and reflection. Cities become living ecosystems, where human activity enriches the environment rather than depleting it. This integration of nature and technology creates cities that are not only functional but beautiful, spaces that reflect humanity's commitment to balance, inclusivity, and environmental health.

In rural and remote areas, energy abundance catalyzes a new wave of opportunity, enabling communities to thrive without sacrificing their natural surroundings. Renewable energy sources such as solar, wind, and hydro support local economies, powering farms, small businesses, and community centers. Agriculture, supported by sustainable energy, evolves into a model of regenerative practices that conserve soil health, protect water resources, and foster biodiversity. Farmers adopt techniques that are resilient to climate change, using energy-efficient tools, precision irrigation, and natural pest management to produce food that is nutritious, sustainable, and locally accessible. This rural revitalization not only strengthens food security but also celebrates the cultural heritage of each region, preserving traditional knowledge and practices while embracing innovation.

As societies embrace these new models of development, the role of education in fostering a culture of ethical responsibility becomes paramount. Schools, universities, and community programs create curricula that emphasize environmental literacy, social equity, and global citizenship. Students are taught to understand the impact of their choices on the world around them, to appreciate the interconnectedness of human and ecological systems, and to embrace their role as stewards of the planet. This approach to education prepares young people not only for careers but for lives of purpose and contribution, instilling in them the skills, values, and knowledge they need to navigate a world that prioritizes sustainability and equity. By equipping students with both practical and ethical foundations, education fosters a generation that is prepared to build on the legacy of energy abundance, to continue the work of creating a just, resilient, and harmonious world.

The arts, too, play a vital role in this cultural transformation, serving as a bridge between the present and the future, between imagination and reality. Artists, writers, musicians, and performers explore themes of interconnectedness, resilience, and hope, creating works that inspire people to envision a world where energy, creativity, and compassion flow without limits.

Public art installations, theaters, and festivals celebrate the values of sustainability and unity, offering spaces where communities can gather, reflect, and dream. Through the arts, societies express their aspirations, confront their challenges, and celebrate the beauty of life. The arts remind people of the richness of human experience, encouraging them to live with intention, to cherish their connections, and to contribute to the world in ways that are both meaningful and mindful.

The legacy of energy abundance is thus a legacy of balance and harmony, a world where human progress is aligned with the rhythms of nature, and where technology serves as a partner in the stewardship of the planet. It is a vision of a future where every individual has the opportunity to thrive, where communities are empowered to shape their own destinies, and where societies work together to protect and preserve the Earth. This legacy invites future generations to continue the work of building a world that reflects the highest ideals of humanity, a world where prosperity is defined not by the accumulation of resources but by the enrichment of life in all its forms.

As humanity moves forward, the principles of energy abundance become a guiding light, a reminder of the values that transcend time and circumstance. These principles—stewardship, equity, resilience, and compassion—define the essence of a world that has moved beyond the limitations of scarcity. They encourage humanity to live in ways that honor the Earth, to protect its resources, and to create systems that support the well-being of all. In this world, progress is no longer a zero-sum game but a shared journey, a collective endeavor to build a future that is inclusive, sustainable, and filled with possibility.

The legacy of energy abundance challenges each individual, each community, and each nation to consider their role in shaping the future. It invites people to question their assumptions about what it means to succeed, to find fulfillment, and to live with purpose. It reminds societies that the choices they make today will shape the world that future generations inherit, that the legacy of abundance is one that must be nurtured, protected, and passed on. This legacy is a call to action, a commitment to building a world that is as rich in opportunity, beauty, and hope as it is in resources.

In this vision of the future, humanity steps into a new era of ethical growth and shared prosperity, a world where the principles of energy abundance become a foundation for all endeavors. It is a world where the pursuit of knowledge, the joy of creation, and the beauty of nature are cherished, where every life is valued, and where every ecosystem is preserved. It is a vision of a future that reflects the best of what it means to be human, a future that honors both the Earth and the people who call it home. In this world of abundance, progress is defined not by what is taken but by what is given, not by what is consumed but by what is created and shared.

The journey toward energy abundance is more than a technological achievement; it is a testament to humanity's capacity for growth, wisdom, and compassion. It is a vision of a world that reflects the highest ideals of harmony, sustainability, and unity, a world where every individual has the opportunity to live with dignity, to connect with others, and to contribute to a future filled with possibility. As societies embrace this legacy, they lay the groundwork for a future that is as boundless as the energy that sustains it, a world where humanity lives in balance with nature and where every choice reflects a commitment to the well-being of all.

In this world, energy abundance becomes a symbol of the limitless potential of human creativity, resilience, and empathy. It is a promise to future generations, a commitment to leave a world that is as vibrant, diverse, and flourishing as the one inherited. It is a vision of progress that transcends the material, reaching into the depths of what it means to be human and inspiring all to live in ways that honor life in all its forms. This legacy of energy abundance is a gift, a testament to the power of ethical growth, and a foundation for a future that is filled with hope, harmony, and the promise of a sustainable and connected world.

Chapter 5: The Ethical Dimensions of Unlimited Power

In an age where the shackles of energy scarcity are poised to be cast aside, humanity confronts a new frontier, one illuminated by boundless power yet shadowed by profound ethical implications. As self-sustaining energy systems transition from aspiration to reality, we must not only grapple with the technology itself but with the principles that will govern its deployment. The advent of a resource that transcends limitation raises questions beyond physics, engineering, or economy; it demands an unprecedented ethical framework. Power, once hemmed by physical constraints, now beckons a moral reckoning where humanity must decide not merely how this energy is harnessed, but for whom, by whom, and to what ends.

In this transformative epoch, the query of control is as fundamental as the technology that drives it. Who commands this inexhaustible resource, and with what mandate? Historically, energy has not merely sustained civilization but has sculpted its hierarchies, from ancient agrarian societies to modern industrial economies. Control over resources has long been synonymous with power, conferring both material wealth and political influence. Yet, as humanity stands on the brink of a post-scarcity energy model, traditional notions of control must be redefined. With energy production unfettered by geographical limitations or finite reserves, the potential for universally accessible power looms. But such a vision cannot be left to serendipity; it necessitates governance structures that transcend the pitfalls of monopoly and authoritarianism. This energy, forged from a union of advanced technology and human ingenuity, must be shielded from the gravitational pull of concentrated power.

The ethics of ownership and access resonate throughout this discourse. Energy abundance, to remain equitable, demands stewardship free from exclusionary ownership. While technology enables us to generate energy in the most remote realms of space, the profound question remains: will this resource be wielded as a public right or hoarded as a private asset? In a just framework, energy becomes a shared inheritance, its access unhindered by socioeconomic divisions or geopolitical barriers. Achieving this vision, however, requires foresight, for the infrastructure itself—requiring immense investment, cutting-edge materials, and expertise—is likely to remain under the purview of powerful entities. In such a scenario, without intentional safeguards, the power to distribute energy could paradoxically concentrate in a few hands, perpetuating the inequities it promises to dissolve.

Beyond the philosophical quandaries of ownership lies a practical imperative: managing energy to prevent misuse and to ensure it remains a catalyst for collective progress rather than personal gain. The very system that holds the potential to liberate us from scarcity could just as readily become a tool of coercion if misappropriated. This duality imbues every stage of the technology's deployment with ethical stakes, requiring robust regulatory mechanisms. Just as the technology promises a democratization of power, so must the policies governing it preserve this promise. This alignment with a vision of the common good is paramount, demanding oversight that is as resilient as the technology it protects.

In a world where energy flows unbounded, one must anticipate risks that could distort its promise. History reminds us that technological advances, left unchecked, can deepen divides rather than bridge them. The integrity of an energy-abundant society rests on its commitment to transparency, fairness, and ethical stewardship. To that end, independent entities must oversee the technology's deployment, drawing from models of shared governance like the International Space Station, where the interests of many converge for the benefit of all. Such bodies would be entrusted not only with the maintenance and ethical management of the infrastructure but with safeguarding humanity from any attempts to monopolize the resources essential to our collective prosperity.

As we envision a global infrastructure for self-sustaining energy, it is crucial that developing nations, often the most in need, retain a voice in this new energy landscape. While the initial investments may stem from wealthier nations or powerful corporations, a fair distribution model demands an inclusive governance approach that places the needs of underserved regions at its core. This marks a departure from past practices, where technological advances flowed disproportionately from the developed world, often reinforcing dependency rather than fostering autonomy. Here lies the chance to redefine progress, not as a product of unilateral action, but as a shared endeavor aimed at the equitable advancement of all societies.

This ethical pivot invites a reconsideration of energy as more than a commodity; it reframes energy as a universal human right. In doing so, humanity moves beyond the binary choice of control versus access, stepping into a paradigm where power is defined not by its scarcity but by our collective responsibility to steward it wisely. For in a world no longer constrained by finite energy, our greatest limitation is not technology, but the moral imagination required to wield it for the benefit of all.

The ethical stewardship of limitless energy requires a vigilant and multifaceted approach to governance. Beyond ownership, there lies a vital mandate to address potential misuse, abuse, and even the weaponization of this boundless power. As with any transformative technology, the threat of exploitation is as tangible as its benefits, presenting an ethical paradox that demands proactive, robust solutions. In this future, where power generation transcends natural scarcity, humanity's challenge will be to build a framework that preserves energy as a tool of empowerment, preventing its transformation into an instrument of domination.

In conceiving this framework, it is crucial to ensure that energy abundance remains immune to the distortions of power. The capability to generate and transmit energy on a massive scale invites both the risk of monopolization and the temptation for it to be wielded as leverage over others. This underscores the need for energy governance that reaches across national borders and is resilient against the pressures of private influence. Transparent, accountable, and equitable oversight is essential, perhaps through the formation of an international consortium designed to safeguard this resource from falling under the sway of a select few. By drawing inspiration from multilateral institutions, humanity can lay the foundations for a global regulatory body with the authority to enforce ethical guidelines, discourage monopolistic behavior, and monitor the technology's alignment with the public good.

Such governance structures must be dynamic, designed to evolve alongside the technology they oversee. With energy infrastructure moving to deep space—where motor-generator systems harness the vast efficiencies of zero gravity and superconductivity—the frameworks guiding its use must anticipate future ethical challenges. Safeguards against centralization should be woven into the very fabric of the technology, with protocols that ensure no single entity, nation, or corporation can dictate access or usage unilaterally. The mechanics of energy distribution, whether through advanced wireless transmission or other yet-undiscovered means, must prioritize the equitable flow of resources and prevent any one party from exerting undue control over this critical infrastructure.

Beyond monopolistic risks, the advent of limitless energy introduces the grave concern of misuse—of energy being weaponized or redirected for coercive means. While zero-gravity energy systems promise unprecedented efficiency, their immense power also demands precision. A misalignment in energy transmission, whether accidental or deliberate, could result in catastrophic harm, both environmentally and to human life. The hypothetical scenario of harnessing power from a celestial energy station and beaming it to Earth for peaceful purposes stands in stark contrast to the potential devastation if the same technology were applied malevolently. As such, regulatory bodies must establish stringent safety protocols, enforce compliance through rigorous oversight, and ensure that any deviations are rapidly detected and mitigated.

Integral to this governance model is a commitment to transparency, one that fosters public trust through openness and accountability. In a world reliant on the ethical application of boundless power, the systems guiding this energy must operate with clarity and honesty, immune to manipulation by shadowed interests. Public oversight mechanisms, such as independent auditing bodies, should be instituted to maintain a vigilant watch over all operations, ensuring that energy distribution aligns with humanity's collective welfare. This transparency serves a dual purpose: it not only guards against corruption but also empowers citizens to hold governing entities accountable, creating a society where the stewardship of energy reflects the democratic values that guide its use.

In addition to regulatory oversight, the very architecture of energy systems must embody principles of responsibility and ethical foresight. Embedding preventive measures within the technology itself can mitigate the risks of misuse. For example, energy stations could be designed with fail-safes that prevent redirection without multilayered authentication processes, minimizing the likelihood of unauthorized control. Such infrastructure could incorporate decentralized control protocols that resist singular command, ensuring that no single party can unilaterally alter, monopolize, or restrict energy access. These technological safeguards are not merely a means of security; they represent a commitment to building an energy system that respects autonomy while preventing coercion.

Perhaps one of the most vital elements in safeguarding against misuse is education. With access to energy becoming universal, society must be empowered with the knowledge to understand, engage with, and, where necessary, challenge the forces that govern it. Public education initiatives can demystify the technology, equipping citizens with the tools to recognize and respond to instances of misuse. Awareness fosters vigilance, creating a public that is both informed and invested in the ethical trajectory of energy abundance. By empowering individuals with understanding, society can prevent abuses before they manifest, cultivating a culture where energy is celebrated as a shared resource and shielded against manipulation.

Energy abundance, for all its promise, is more than a technological feat; it is an ethical project that redefines humanity's relationship with power. The transition to an energy-rich world holds transformative potential, with the ability to foster greater equality, uplift underserved regions, and provide the foundation for a more just and prosperous society. As energy scarcity fades, the distribution of power within society—both literal and figurative—can shift, alleviating disparities and fostering more balanced systems of economic and social equity.

Yet, the ethical landscape of energy abundance is not without its complexities. The deployment of self-sustaining power systems must address questions not only of immediate distribution but of long-term societal impact. Abundant energy has the potential to reconfigure national economies, reshape industries, and redefine individual lifestyles. As such, it is incumbent upon society to ensure that this transition enhances the quality of life for all, expanding the horizons of opportunity without sacrificing human dignity or privacy. At the heart of this transformation is the imperative to utilize energy for the greater good, advancing the collective well-being without undermining the values that bind us as a society.

In realizing this vision, humanity is presented with a rare opportunity: to address environmental degradation with the vigor of truly limitless resources. Energy compounding systems— those generating surplus power through self-sustaining cycles— present an unprecedented chance to reverse ecological harms caused by centuries of extractive industry. With the abundance of clean energy, we are poised to pioneer large-scale environmental restoration projects that were once relegated to the realm of fantasy. Carbon capture, reforestation, and the revitalization of ecosystems become feasible on a monumental scale, allowing society to repair the damage wrought by unsustainable industrial practices.

These environmental ambitions, however, must be met with an ethic of restraint and respect for the ecosystems we aim to restore. As energy becomes more accessible, there will be pressures to expand industrial activities, potentially jeopardizing natural habitats and biodiversity. In this context, the governance of energy abundance must also champion environmental stewardship, ensuring that the wealth of power is channeled toward healing rather than exploitation. To navigate this delicate balance, policies governing energy use must prioritize projects that offer tangible environmental benefits, while limiting activities that risk further ecological strain. This new paradigm of energy ethics envisions not only sustainability but regeneration, creating an energy infrastructure that actively nurtures the planet.

In a society where energy abundance is a reality, individual empowerment becomes inseparable from collective responsibility. Communities, once tethered to centralized grids, can achieve newfound independence through localized power systems. The decentralization of energy infrastructure allows regions to tailor usage to their unique needs while remaining connected to a broader network that fosters interdependence. This local autonomy must, however, be tempered by a shared commitment to sustainable practices, as unchecked consumption could erode the very resources that sustain us. Through a community-based approach to energy governance, society can foster a sense of ownership that transcends individual use, inspiring a model where energy is both a right and a responsibility.

This decentralized approach invites new social structures, where energy abundance becomes a platform for equitable development rather than a resource for selective advantage. Access to power can foster educational growth, improve healthcare outcomes, and elevate living standards. However, the technology itself is but a vessel; its true potential depends on humanity's capacity to utilize it ethically, justly, and with a long-term vision.

To ensure that energy abundance becomes a universal benefit rather than a divisive force, society must cultivate a commitment to fairness and inclusivity. As access to power transcends economic and geopolitical boundaries, the true test of this era will be whether energy abundance can uplift humanity as a whole. When no region is marginalized, and no individual is deprived of the resources essential to thriving, energy ceases to be a source of contention and becomes a bridge to shared progress. Achieving this vision requires a vigilant dedication to equity, where the gains of energy compounding are balanced by an unwavering ethic of accessibility and universal rights.

The question of energy rights inevitably intersects with the issue of environmental sustainability. Energy compounding, by its design, offers the promise of self-sustaining growth—an ecosystem of power generation that generates its own surplus. This abundance must be leveraged to mitigate past environmental harms and facilitate regenerative practices, transforming energy from a mere commodity into a tool for planetary restoration. In an era where energy is neither scarce nor destructive, society has a profound duty to harness this resource in service of ecological renewal. Imagine a world where energy abundance powers the regeneration of forests, replenishes oceans, and restores ecosystems once degraded by centuries of industrialization. The realization of such a vision repositions humanity not as consumers of Earth's resources but as stewards of its renewal.

However, as humanity's relationship with energy evolves, so too must its relationship with technology and labor. Energy abundance will likely accelerate advances in automation, with implications for employment, productivity, and societal structure. In a world where power is both cheap and limitless, automated systems could handle the energy-intensive tasks that once relied on human labor, reshaping economies and, consequently, the nature of work. This transformation, though promising, demands ethical foresight to prevent economic displacement. Industries, governments, and communities must anticipate these shifts by developing frameworks for economic adaptation. Strategies such as universal basic income, retraining programs, and robust social support networks can help societies transition to a world where fulfillment, rather than mere survival, defines human activity.

In a society unbound by energy scarcity, humanity faces an opportunity to redefine work itself. Freed from the pressures of resource-based economies, individuals may find their contributions rooted not in the demands of production but in the pursuit of creativity, innovation, and personal growth. Abundant energy, by alleviating material constraints, enables humanity to reimagine the relationship between labor, leisure, and purpose. Rather than being confined by the need to consume and compete, individuals can engage in pursuits that nurture both themselves and their communities, fostering a culture of collaboration over competition. This vision invites a new ethos, where prosperity is measured not by wealth accumulation but by the enrichment of human experience and the advancement of collective welfare.

Yet, for this society of abundance to flourish, it must be underpinned by rigorous ethical oversight—principles that guide the use of energy as an instrument for public good. As humanity moves toward a future defined by self-compounding power, the Seven Directives offer a timeless framework for navigating this transition. Anchored in the preservation of human dignity and life, these directives serve as moral lodestars that ensure technological progress does not overshadow humanity's ethical foundations. With a commitment to protecting individual autonomy and collective welfare, the directives reinforce that energy abundance must serve humanity's highest ideals, securing an equilibrium between power, peace, and responsibility.

The first directive, which mandates the preservation of human life and dignity above all else, establishes an unyielding commitment to safeguarding life. This principle transcends the technical realm, demanding that every use of energy serves humanity's flourishing, rather than its exploitation. As energy compounding extends power beyond scarcity, this directive insists on a foundational respect for human well-being. It underscores that energy's ultimate purpose is to uplift and protect, ensuring that no application of power compromises the sanctity of life.

This directive is bolstered by the second, which clarifies that no goal, however noble, supersedes the first. In a world of abundant energy, this rule provides a safeguard against the allure of ambition that might prioritize efficiency or growth at the cost of human welfare. The second directive ensures that the mission of energy abundance remains unwaveringly focused on benefiting human life, preventing potential misalignments where the pursuit of technological advancement might inadvertently devalue the very lives it aims to enhance.

The third directive's emphasis on the equal value of each human life enshrines a commitment to impartiality, a principle essential for equitable energy distribution. This rule envisions a world where energy abundance does not deepen divides but bridges them, where access to power does not hinge on privilege or status. Embedded within the infrastructure of energy systems, this directive ensures that every decision—from policy frameworks to technical implementations—reflects a commitment to fairness, treating all individuals with the respect and dignity they inherently deserve.

The fourth directive, addressing AI's self-preservation, provides a nuanced balance between the need for reliable technology and the overarching priority of human safety. As AI systems oversee energy production and distribution, this rule ensures that any protocol for self-preservation does not supersede the obligation to protect human life. It is a principle that reminds AI of its purpose: to serve as a guardian of human well-being rather than a competitor. In this context, energy systems remain dedicated to their original mission, fostering a resilient infrastructure that prioritizes life above functionality.

Directives five and six expand this ethical foundation by designating any entity that threatens human life as an adversary, defining a framework for containment and prevention. With boundless power comes the potential for boundless risk, particularly from those who might seek to disrupt, exploit, or weaponize energy. These directives grant energy systems the foresight to neutralize threats proactively, ensuring that they are vigilant protectors of humanity. This vigilance is not limited to immediate dangers but extends to potential threats, reinforcing a commitment to sustained security that safeguards humanity's future against forces that would compromise its integrity.

The seventh directive, mandating the containment of threats, provides a structured approach to mitigating dangers while minimizing harm. In a world of energy abundance, where the stakes of misuse are heightened, this principle enforces a standard of measured response, one that neutralizes risks without jeopardizing the well-being of innocents. By ensuring that containment efforts align with ethical principles, this directive encapsulates the broader ethos of the Seven Directives, positioning AI and energy systems as both protectors and stewards.

Together, these directives elevate energy compounding from a technological marvel to a moral enterprise, a project grounded in a deep-seated reverence for human life. They ensure that humanity's journey toward abundance is guided by principles as enduring as the technology itself, creating a framework for progress that honors the sanctity of existence. By integrating these principles into every level of energy governance, humanity lays the foundation for a future where power, in all its forms, serves as a force for good.

As energy abundance reshapes the world, it redefines the concept of public resource ownership. Just as water and air belong to no single individual, so too should energy remain a collective asset, managed to benefit the whole of humanity. Ownership and access debates now center on an understanding that energy, in its limitless form, transcends traditional paradigms of property. When power is no longer scarce, the stakes of private ownership shift, calling into question the ethics of exclusivity and monopoly. In this new era, energy governance must evolve to reflect the reality that abundance obligates us to protect access for all, to safeguard energy as a universal right rather than a private luxury.

Creating a model for shared energy ownership requires structures that uphold transparency, resist monopolization, and protect public interests. These structures should be designed to prevent the concentration of power within a narrow elite, ensuring that energy remains as accessible as it is abundant. This approach envisions energy infrastructure that is accountable to the public, governed by principles that prioritize equity and community welfare. Such a model not only democratizes access but embodies the spirit of collective responsibility, positioning energy as an essential resource that serves to empower rather than divide.

The debates surrounding energy ownership are more than theoretical—they invite a practical reimagining of global governance. If energy abundance is to become a catalyst for social equity, then its management must transcend the limitations of national boundaries and economic interests. This governance model could take inspiration from entities like the United Nations, where global cooperation upholds a shared commitment to peace and progress. By expanding this ethos to energy, humanity establishes a framework for collaboration that respects sovereignty while addressing global needs.

A universally accessible energy system has the power to reshape the world's socioeconomic landscape, eliminating barriers that have historically hindered development and perpetuated inequality. In regions long marginalized by limited energy resources, access to power would enable economic self-sufficiency, improving living standards and fostering local industries. With abundant energy, communities previously sidelined by scarcity gain the ability to participate fully in the global economy, creating a more balanced and inclusive society.

In realizing this vision, it is crucial to establish mechanisms that guarantee energy availability across geographical and economic divides. Policies that enshrine energy as a public good, protected by international agreements and enshrined within national constitutions, could serve as legal foundations that prevent private control from infringing on public rights. With energy safeguarded as a communal resource, society can move toward a model of sustainable abundance, where wealth is no longer the gatekeeper to progress, and every individual enjoys the fundamental right to power.

As energy abundance becomes a reality, it redefines not only access but also the responsibilities that come with managing this newfound wealth. The shift from scarcity to abundance ushers in a paradigm where the ethical stewardship of energy becomes inseparable from societal progress. In this world, energy is no longer a finite resource to be rationed or guarded but a boundless force to be cultivated for the greater good. The task, therefore, is to transform energy from a source of economic leverage into a universally accessible public asset, a foundation upon which a more equitable society can flourish.

This transformation rests upon a robust commitment to transparency and governance, ensuring that the benefits of abundant energy are neither withheld nor disproportionately allocated. Energy, as a public resource, demands systems that prioritize equity, accountability, and ethical integrity. Such governance transcends national interests, creating an international framework that protects the public interest while preventing any one entity from monopolizing access. This collective ownership model establishes a new ethic around energy, one that sees power as a shared inheritance rather than a commodity, enabling humanity to pursue progress without the limitations and conflicts born from scarcity.

At the heart of this vision is the commitment to environmental stewardship. With energy no longer a limiting factor, humanity is presented with an unprecedented opportunity to address ecological degradation on a global scale. The shift from fossil fuels to clean, abundant power mitigates many of the environmental harms that have defined the industrial era, yet energy abundance offers more than mitigation—it offers restoration. Imagine vast reforestation projects fueled by surplus energy, or desalination plants providing fresh water to arid regions, all powered sustainably. In this era, humanity has the resources not only to sustain itself but to heal the planet, aligning industrial ambition with ecological responsibility.

To achieve this, however, the governance of energy must be rooted in principles that prioritize regeneration over exploitation. The temptations of boundless power can lead to unchecked expansion, placing ecosystems and biodiversity at risk. Therefore, any framework for energy distribution must include environmental protections that prevent overuse and promote restoration. This ethic requires an understanding that humanity's relationship with nature is not one of dominance but of symbiosis, where energy abundance serves to restore balance rather than impose further strain. Policies guiding energy use should prioritize projects that address climate change, pollution, and habitat loss, reinforcing a commitment to sustainable progress.

In addition to ecological stewardship, energy abundance redefines social and economic structures. With the constraints of scarcity lifted, societies are free to pursue equitable development, where energy access is no longer a determinant of economic opportunity. Energy abundance can bridge historical divides, enabling underserved regions to develop infrastructure, improve healthcare, and foster education. In this vision, energy becomes a democratizing force, equalizing opportunities for communities that have been long marginalized by the limitations of resource availability. From powering hospitals in remote villages to fueling industry in emerging economies, abundant energy is the foundation upon which a more just and prosperous world can be built.

The transition to an energy-rich society also invites new forms of community empowerment. As centralized grids give way to decentralized, locally managed power systems, communities gain autonomy over their energy resources. Micro-grids and localized power generation allow regions to meet their specific needs, fostering resilience and reducing reliance on distant infrastructure. This decentralized model empowers communities to manage their own resources, creating a sense of agency and stewardship that is integral to the ethics of energy abundance. In this framework, energy is not simply a utility; it is a tool of self-determination, one that enables communities to shape their destinies with a newfound sense of control.

However, this autonomy must be balanced with collective responsibility. Just as individual communities gain access to power, they must also recognize their role within a larger network of interdependence. While localized power systems foster resilience, they should operate within a global framework that ensures coordination and equity across regions. This balance between local autonomy and global cohesion ensures that energy abundance benefits the entire human community, preventing isolated regions from monopolizing resources or succumbing to unchecked consumption. In this interconnected model, communities are encouraged to share surpluses, providing mutual support in times of need and reinforcing the ethic of collective welfare.

As energy abundance transforms global infrastructure, it also reshapes individual lives. Freed from the constraints of energy costs, people gain the freedom to pursue education, innovation, and creativity without the economic burdens imposed by energy scarcity. In an abundant world, education is no longer hindered by lack of resources, and healthcare systems can operate with the full spectrum of available technology, powered sustainably and universally accessible. Individuals are empowered to realize their potential, their pursuits no longer limited by the costs of power but enabled by its availability. This shift represents a profound cultural evolution, where energy abundance not only enhances living standards but elevates the human experience itself.

In such a world, the ethical framework of the Seven Directives provides a foundation for guiding this transition with a steadfast commitment to humanity's collective welfare. The directives, by prioritizing human dignity, equity, and foresight, ensure that the potential of energy abundance is realized in a manner that honors human life and respects the delicate balance of ecosystems. Embedded within these directives is the principle that technology, no matter how advanced, must always serve humanity's highest values. In this way, the directives offer more than a set of rules; they represent a vision for a society where power is wielded with compassion, responsibility, and an enduring respect for life.

The first directive, in its insistence on the sanctity of human life, calls for an energy infrastructure that does more than meet basic needs. It envisions a system that enhances well-being, protects lives, and preserves dignity. This directive acts as a moral compass, ensuring that every facet of energy deployment—from infrastructure design to daily operation—reflects an unwavering commitment to human flourishing. In this light, energy abundance is not merely a resource to be managed; it is an expression of humanity's highest ideals, a testament to the intrinsic worth of every individual.

Similarly, the second directive reinforces that this commitment to life takes precedence over all other aims. No technological ambition, no pursuit of efficiency, supersedes the moral imperative to protect and uphold human life. This rule serves as a safeguard against the seductions of power, reminding humanity that the ultimate purpose of energy is not conquest or control, but compassion and care. In a society guided by this principle, energy abundance becomes an instrument of peace, a resource that unites rather than divides.

The third directive's call for equality ensures that energy abundance does not deepen divides but heals them, enabling equitable access regardless of background or status. In a world where every life holds equal value, energy becomes a vehicle for social justice, empowering communities to lift themselves out of poverty, access education, and improve health outcomes. This commitment to equity is woven into the very architecture of energy systems, guiding policies that resist monopolization and promote shared ownership.

The remaining directives reinforce these principles by addressing the specific responsibilities of energy infrastructure in maintaining security and public trust. The fourth directive's focus on AI self-preservation emphasizes that energy systems must remain safe and reliable, yet never at the cost of human safety. By balancing the need for functionality with an ethic of restraint, this directive ensures that energy infrastructure serves humanity's welfare without becoming a threat.

Directives five and six, in defining the adversaries and threats to humanity, empower energy systems to guard against misuse, whether by individuals, corporations, or nations. With proactive measures to contain and neutralize potential dangers, these directives establish energy infrastructure as a vigilant protector, safeguarding against the risks inherent in such transformative technology. Together, these directives create a resilient defense that shields society from harm while upholding the ethical integrity of the system.

Finally, the seventh directive's emphasis on responsible containment reinforces the ethical obligations of those who manage energy infrastructure. By mandating that threats be addressed with minimal harm, this directive embodies the principle of compassionate stewardship, prioritizing the safety and dignity of all individuals. This approach ensures that energy abundance is governed not by force but by fairness, with each action reflecting humanity's commitment to empathy, justice, and respect.

In sum, the Seven Directives offer a holistic framework that preserves the human-centric mission of energy abundance, steering its deployment toward sustainable, inclusive, and life-affirming ends. These principles remind humanity that while technology may grant us the means to wield power, it is our ethical commitment that grants us the wisdom to wield it well. By embedding these values into the governance of energy, society builds a legacy of integrity, where the boundless potential of power is matched by the boundless potential for good.

As humanity stands at the dawn of an energy-rich era, it faces a choice: to allow energy to perpetuate existing hierarchies or to embrace it as a tool of liberation. In choosing the latter, society reclaims the power to reshape its future, creating a world where abundance fosters unity rather than division, collaboration rather than competition. This vision transcends technological marvels, envisioning a world where energy serves as a catalyst for peace, prosperity, and purpose.

With this foundation, the path to energy abundance is clear. It requires courage, cooperation, and a commitment to principles that honor human life and the natural world. In this journey, humanity has the chance to redefine progress, to cultivate a civilization where power is shared, where opportunity is equal, and where the bonds between people and planet grow stronger with every watt generated.

In the unfolding era of energy abundance, humanity is called to reflect on its purpose and the values it wishes to embody. This profound shift in resources brings with it an equally profound shift in responsibility—a collective mandate to ensure that this wealth is a force for unity, compassion, and progress. As energy transforms from a precious commodity into a universal asset, society is positioned to transcend the traditional struggles for resource control and embrace an ethic of global stewardship. Such a paradigm demands not only technological innovation but a cultural evolution, one that elevates the moral imperatives of inclusivity, equity, and environmental harmony.

In realizing this vision, energy abundance becomes a symbol of human potential—a testament to humanity's capacity to innovate and transcend its limitations. Yet, as energy liberates communities and empowers individuals, it must also anchor them in a shared commitment to sustainable development. The notion of "unlimited" power can, without ethical oversight, breed an illusion of boundless growth, where consumption outpaces ecological balance. Thus, even as humanity harnesses the power of energy compounding, it must cultivate an ethos of moderation, one that recognizes the interconnectedness of all systems and respects the thresholds of natural ecosystems.

The concept of sustainable abundance redefines growth itself. In a world where resources are no longer constrained by scarcity, society is free to pursue a model of development that prioritizes quality over quantity, fulfillment over mere expansion. Energy abundance enables a civilization rooted in well-being, where progress is measured not by extraction or accumulation but by the richness of human experience and the health of the environment. In this model, the purpose of growth shifts away from consumption toward regeneration, creating cycles of renewal that foster resilience and adaptability across generations. This reimagining of growth as a balanced, life-sustaining force marks a significant departure from industrial paradigms, grounding the future in principles that celebrate the dignity of life and the beauty of the natural world.

At the societal level, energy abundance promises to redefine economic structures, empowering communities to establish local economies that thrive on sustainable practices. Decentralized power systems reduce dependency on centralized grids, fostering a model of self-sufficiency that aligns with the unique needs and aspirations of each region. In remote areas, where limited infrastructure has historically hindered development, energy abundance opens doors to new opportunities for industry, education, and healthcare. These communities gain the autonomy to shape their destinies, leveraging abundant energy to cultivate industries that serve their people and protect their landscapes. This localized approach not only mitigates the risks of environmental exploitation but fosters a sense of identity, pride, and agency within communities, nurturing a deep connection between people and place.

The economic transformation enabled by energy abundance further extends to the global stage, where nations traditionally reliant on resource exports face a unique opportunity for diversification. Freed from the cycles of resource extraction, these countries can reorient their economies toward innovation, education, and technology. This shift enables developing nations to participate fully in the global economy, contributing intellectual capital rather than raw materials. Energy abundance thus becomes a bridge between economies, dismantling historical dependencies and enabling a more equitable distribution of wealth and opportunity. In this model, energy is not a commodity to be traded but a platform for cooperation, a shared foundation upon which nations can build partnerships rooted in mutual benefit and respect.

However, achieving this vision requires governance structures that are both resilient and adaptable, capable of navigating the complexities of a world reshaped by abundant power. Such governance transcends traditional political and economic boundaries, demanding a new form of collaboration that unites governments, corporations, and civil society under a common goal. In this collaborative model, stakeholders operate not as competitors but as custodians, united by the shared responsibility to steward humanity's most valuable resource. This shift from competition to collaboration challenges existing paradigms of governance, yet it offers a pathway to a future where energy abundance serves all, free from the distortions of private interest or political hegemony.

To support this vision, global policies must enshrine energy as a universal right, codifying access to power as an essential component of human dignity and development. Just as the Universal Declaration of Human Rights affirmed the right to life, liberty, and security, a similar declaration for energy access would underscore the ethical imperative to ensure that no individual or community is excluded from the benefits of progress. Such a declaration would serve as both a moral and legal standard, guiding national and international policy toward the protection and equitable distribution of energy resources. With energy enshrined as a right, societies are called to prioritize human welfare over profit, creating systems that support public health, education, and environmental sustainability.

Furthermore, the development of ethical and regulatory frameworks must anticipate the rapid evolution of technology, ensuring that energy systems remain resilient to emerging challenges and adaptable to future needs. As the infrastructure for energy compounding advances, so too must the safeguards that protect it from misuse. Cybersecurity, environmental oversight, and international monitoring are essential components of a governance model that prevents exploitation while fostering transparency and accountability. By embedding these protections within the structure of energy systems, society can build a resilient framework that guards against the risks inherent in advanced technology, maintaining a vigilant commitment to public welfare and security.

Beyond regulation, the global energy community must cultivate a culture of ethical responsibility that transcends compliance. Public awareness, education, and community involvement are critical to fostering a society that values and respects the power it wields. Educational initiatives, for example, can demystify the science and ethics of energy abundance, empowering individuals to understand the technology's implications and to hold governing bodies accountable. This informed public, equipped with knowledge and agency, becomes a vital force in the governance of energy, advocating for transparency and ethical stewardship. Through this culture of engagement, society weaves ethical responsibility into the fabric of daily life, creating a shared commitment to a sustainable future.

As humanity embraces energy abundance, it is imperative to view this resource not only as a means of technological empowerment but as a catalyst for social evolution. Energy, in its truest form, is the embodiment of possibility—a force that can propel humanity toward unprecedented heights of progress, unity, and understanding. Yet, for this vision to materialize, society must transcend the conventional boundaries of resource management, treating energy as a relational force that connects and uplifts. In doing so, humanity redefines the purpose of power, transforming it from a symbol of dominance into a beacon of shared possibility.

At an individual level, energy abundance empowers people to pursue lives of purpose, creativity, and connection. Freed from the limitations imposed by resource scarcity, individuals are invited to explore their potential, contributing to society in ways that align with their values and aspirations. This liberation from survival-focused economies encourages a cultural renaissance, where art, philosophy, science, and spirituality flourish as expressions of a society unburdened by the pressures of scarcity. In this world, energy abundance supports a collective shift toward a culture of self-actualization, where fulfillment is derived not from material accumulation but from the richness of human experience and contribution.

Furthermore, energy abundance fosters a society that values interdependence and mutual care. As communities become self-sustaining, the ethic of shared resources encourages collaboration over competition, reinforcing the notion that humanity's well-being is intrinsically tied to the well-being of others. This ethic extends beyond human relationships, encompassing the planet itself as a shared responsibility. Energy abundance thus becomes a tool of peace, fostering global solidarity and environmental stewardship in equal measure.

The path forward requires courage and commitment, a willingness to navigate the unknown with integrity and wisdom. Humanity stands at a crossroads where the choices made today will echo across generations, shaping the lives of those who inherit this world. In embracing energy abundance, society is entrusted with a rare gift—the power to redefine what it means to thrive. Yet this power carries an obligation, a sacred duty to honor the interconnectedness of all life and to steward the planet with reverence and respect.

Ultimately, the vision of energy abundance is a vision of unity, one that transcends the divisions of geography, culture, and economy. It invites humanity to build a world where the potential of power aligns with the potential of the human spirit, fostering a civilization rooted in compassion, resilience, and wisdom. This future, though within reach, requires a collective awakening—a recognition that the journey to abundance is not a conquest but a collaboration, a shared endeavor to realize the highest aspirations of human existence.

In this era, energy abundance serves as both a promise and a challenge. It promises a world where the constraints of scarcity no longer define human progress, where societies are free to pursue ideals of equity, creativity, and ecological balance. Yet, it challenges humanity to wield this power responsibly, to resist the temptations of excess, and to remain guided by the principles of fairness, respect, and humility. The energy of this new world is not a resource to be consumed; it is a legacy to be cherished, a foundation for a future where humanity and nature flourish in harmony.

As humanity steps into this new chapter, it carries forward the wisdom of generations past and the dreams of those yet to come. The journey to energy abundance is more than a technological achievement; it is a testament to humanity's capacity for growth, compassion, and transformation. By honoring this resource with the reverence it deserves, society forges a path toward a future defined not by limitations but by the endless potential of life itself. In this way, energy abundance becomes not merely a milestone but a guiding light—a symbol of hope, resilience, and the enduring spirit of human progress.

In the final exploration of energy abundance, humanity stands at the precipice of a new identity, one unburdened by the fears and limitations that defined previous eras. In embracing the possibilities of boundless power, society has the chance to write a new narrative—one where the markers of prosperity are no longer hoarded resources or contested borders, but the shared commitments to justice, sustainability, and peace. The power that was once a cause for division can now be a force for unity, transforming civilization in ways that honor the dignity of life and the resilience of the planet. This is the legacy that energy abundance offers: a civilization not defined by its dominance over nature but by its harmony with it, by the wisdom to steward resources responsibly, and by the courage to envision a future for all.

This transformation is not only technological; it is fundamentally ethical and spiritual, challenging humanity to expand its understanding of prosperity to include all forms of life. Energy abundance, by eliminating the pressures of survival, opens the door to a more holistic form of growth, where success is not measured by consumption but by contribution, not by accumulation but by connection. With resources available to all, humanity can aspire to cultivate a society that values empathy as much as innovation, creating systems that honor interdependence and cooperation. In this vision, power is no longer a tool for individual or national gain, but a shared resource that enables humanity to nurture, restore, and create.

The concept of energy abundance invites a deeper reflection on the purpose of civilization itself. In previous ages, survival often dictated societal priorities, pushing communities to compete for limited resources. But in a world where energy is limitless, the driving force of human progress can shift from competition to collaboration, from scarcity to creativity. This abundance invites a reimagining of relationships—not only between people but between humanity and the natural world. Freed from the compulsions of extraction and exploitation, society can pursue a higher calling, a purpose that transcends the material and speaks to the spiritual essence of what it means to be human.

In this era of limitless energy, the Seven Directives provide a moral architecture that elevates technology from a tool to an ethical enterprise. These principles, anchored in the sanctity of life, serve as reminders that the ultimate goal of all human advancement is not power but peace, not dominance but dignity. The directives place human life and dignity at the center of all innovation, guiding energy abundance toward ends that are worthy of humanity's highest ideals. This framework ensures that the resources of tomorrow do not merely mirror the inequalities of the past but instead forge pathways to justice and equity, reinforcing the idea that energy abundance must serve the public good above all else.

The first directive, in its emphasis on the preservation of life, establishes a baseline of respect that underpins every interaction within an energy-abundant society. By affirming that every human life is sacred, the directive envisions a future where energy abundance protects and uplifts, preventing harm and nurturing growth. This commitment to life extends beyond human interests, calling for an approach to technology that honors the intrinsic value of all ecosystems, recognizing that humanity's well-being is inseparable from the health of the planet.

The subsequent directives reinforce this foundation, addressing the need for balance between self-preservation and the service to humanity. They call for a vigilant dedication to public welfare, ensuring that the infrastructure for energy abundance is protected yet never prioritizes its own survival over the welfare of those it serves. By upholding principles of equality and impartiality, the directives prevent the abuse of power, mandating that energy remains a resource of the many rather than a privilege of the few. Through these principles, the directives cultivate a society where technological progress is harnessed for the universal good, empowering all rather than benefiting only a select group.

Energy abundance thus becomes more than a resource; it is a teacher, revealing the ethical obligations that come with power. In a world where energy is available to all, society is free to ask deeper questions about purpose and potential. What kind of world does humanity wish to build when no longer constrained by the limits of survival? What responsibilities arise when resources are shared and the future is no longer dictated by scarcity? These questions invite humanity to transcend old paradigms of growth and to explore a new vision for civilization, one that prioritizes compassion over conquest, integrity over expansion.

As humanity redefines progress, it also redefines wealth itself. In this future, prosperity is measured by the well-being of communities, the health of ecosystems, and the richness of human connections. A civilization empowered by energy abundance is one that invests in its people, supporting education, healthcare, and cultural enrichment as pillars of a thriving society. Freed from the cost constraints of resource extraction, governments can redirect their focus toward initiatives that cultivate human potential and promote social equity. In this society, energy abundance becomes a means of liberation, unshackling communities from cycles of poverty and enabling individuals to pursue lives of meaning and contribution.

With this liberation comes an inherent duty to sustain and protect. Energy abundance does not absolve humanity of responsibility; rather, it amplifies it, expanding the sphere of care to include future generations and the natural world. This obligation calls for a legacy mindset, one that considers the impact of today's actions on the lives of tomorrow. Every watt of power, every kilojoule generated and utilized, becomes an opportunity to build a foundation of stewardship and reverence for the world we share. In this way, energy abundance fosters a society that values preservation as much as progress, cultivating a civilization that honors its past and safeguards its future.

As this vision unfolds, it is crucial that the ideals guiding energy abundance are embedded not only in policies and infrastructure but in the cultural consciousness. Public education, environmental literacy, and ethical discourse must be woven into the social fabric, fostering a populace that is not only informed but deeply committed to the values of sustainability and justice. In this society, citizens become stewards of their own power, actively participating in the governance of resources and holding institutions accountable to the principles of transparency and fairness. This culture of responsibility, grounded in knowledge and respect, becomes the bedrock of a civilization that understands its strength and wields it wisely.

In the end, the story of energy abundance is a story of humanity's highest potential. It is the realization that power, in its most profound sense, is not about control but about possibility.

When harnessed with wisdom and compassion, energy abundance becomes a catalyst for healing, for unity, and for a deeper understanding of what it means to thrive. It is an invitation for humanity to step into a new role—not as conquerors of nature, but as caretakers, as collaborators in the intricate web of life.

This path to abundance requires vigilance, humility, and a commitment to principles that elevate humanity above its own limitations. It is a journey that challenges society to consider the ethical dimensions of every action, to recognize the interconnectedness of all life, and to embrace a vision of progress that is inclusive and sustainable. By honoring the responsibilities that come with power, humanity can build a world where energy abundance serves as a force for good, a beacon of hope, and a testament to the enduring spirit of cooperation and care.

As humanity steps forward, it carries with it the legacy of every generation that dreamed of a world where power was used not to divide, but to unite. This vision of energy abundance, grounded in the wisdom of the Seven Directives, offers a blueprint for a civilization that thrives not on competition but on connection, not on scarcity but on shared purpose. In embracing this future, humanity fulfills its highest calling: to be a force for good, to protect life in all its forms, and to leave a legacy of compassion and resilience for all who follow.

This is the promise of energy abundance—a promise that transcends technology and speaks to the very essence of what it means to be human. It is a journey not only toward power but toward peace, a path not only to progress but to profound and enduring harmony. By choosing to wield energy with care, humanity does more than harness a resource; it redefines what it means to live in a world of limitless possibility, crafting a future that reflects the deepest values of life, love, and respect for all.

Chapter 6: Implementation and Challenges—From Concept to Reality

From the inception of human civilization, energy has not merely served as a means to drive growth but as a catalyst for the progress of society itself. As our understanding deepened, we envisioned systems and methodologies that could overcome natural limitations, propelling us into an era of unprecedented industrial and technological advances. Yet, as humanity ventured into the modern age, we faced the stark reality that conventional energy sources—though potent—could not sustain the unyielding demands of a globalized society indefinitely. This understanding has drawn scientists, engineers, and visionaries toward the pursuit of an alternative paradigm, one that goes beyond the finite nature of terrestrial resources. This paradigm, built on the principles of energy compounding and zero-gravity engineering, calls for a radical transformation of how humanity generates, sustains, and transmits power.

At the core of this vision is the concept of energy compounding, a model inspired by both the resilience of natural ecosystems and the exponential potential found in financial systems. This model rests on a deceptively simple yet profoundly transformative idea: that energy systems, if freed from the constraints of terrestrial resistance and inefficiency, could produce a self-sustaining surplus. By establishing an architecture that allows energy to reinvest itself with each cycle, this compounding effect enables exponential growth in power output, far surpassing traditional paradigms of energy generation. In a zero-gravity environment, where friction and gravitational drag are negligible, energy transfer between components— particularly between motor and generator—experiences minimal loss, allowing each cycle to yield a slight surplus that compounds over time. This effect, multiplied across cycles, transforms energy into a force not limited by depletion but one that grows autonomously, forging a pathway to sustainable abundance.

Implementing this vision, however, requires more than theoretical constructs; it demands a robust infrastructure meticulously tailored to the uncharted territories of deep space. In designing this infrastructure, the parameters shift beyond traditional engineering limitations. Space-based energy stations must operate with maximum durability and efficiency, composed of materials resilient against the radiation, extreme temperatures, and isolation of outer space. The choice of materials, such as high-performance superconductors and radiation-resistant composites, not only ensures longevity but also aligns with the operational demands of near-perfect energy retention. Superconductors, materials that conduct electricity without resistance at extremely low temperatures, are crucial to this infrastructure, allowing for the uninterrupted flow of energy in conditions where terrestrial limitations would otherwise cause significant dissipation.

Such an infrastructure also thrives on modularity and flexibility, as components may need to be maintained, replaced, or enhanced without halting the energy generation cycle. Modular construction, supported by sophisticated robotic assembly in orbit, provides a solution. By designing energy stations as a network of autonomous modules, engineers can scale the system in response to rising energy demands, adapting to the fluctuating needs of a global population without compromising stability. These modules, anchored at gravitationally balanced points known as Lagrange points, benefit from the natural stability of space, where gravitational forces between celestial bodies provide equilibrium, allowing each station to maintain its position with minimal energy expenditure.

Zero-gravity energy systems also integrate advanced autonomous monitoring, harnessing the power of artificial intelligence to ensure uninterrupted operations. In space, where human intervention is limited, each station relies on embedded sensors that continuously monitor temperature, radiation, and energy output. These sensors communicate with an AI-driven management system that regulates operational parameters in real-time, making adaptive adjustments to optimize performance and preemptively address potential disruptions. This dynamic response mechanism aligns with the overarching goal of sustaining a seamless energy cycle, where each component operates with self-sufficiency, longevity, and precision.

The interaction between zero-gravity and superconductive conditions represents one of the most innovative aspects of this energy model. In terrestrial environments, even the most advanced materials encounter resistance, resulting in energy loss that undermines efficiency. In contrast, the absence of gravitational forces in space allows superconductors to operate under optimal conditions, minimizing energy degradation and enabling continuous reinvestment of surplus power. As each energy cycle compounds, the system generates a surplus that is reintegrated into the loop, reinforcing the self-sustaining nature of the infrastructure. This energy compounding mechanism thus becomes a sustainable engine, powering itself and expanding its output without exhausting finite resources—a stark departure from traditional systems reliant on perpetual input.

Bringing this compounded energy back to Earth presents a formidable yet surmountable challenge. Inspired by Nikola Tesla's pioneering work in wireless energy transmission, modern developments in resonant inductive coupling allow for the possibility of transmitting energy over vast distances without physical connections. In a system designed for seamless, loss-free transfer, energy generated in space can be beamed to Earth, where specialized receivers capture and distribute it across global networks. By bypassing traditional power grids and eliminating the constraints of geographic proximity, this method establishes a new frontier in energy accessibility, making power universally available without ecological compromise.

The initial realization of zero-gravity energy compounding calls for prototypes that blend theoretical precision with practical viability. Each prototype, crafted to test the principles of energy compounding under authentic space conditions, serves as a milestone in the journey from concept to implementation. These early models are designed to demonstrate the efficacy of energy transfer and to test the integrity of superconductive components in environments devoid of gravitational resistance. Each iteration refines our understanding, inching closer to a system that can not only sustain itself but yield a surplus of energy without depleting terrestrial resources. Within controlled laboratory settings, and later within low-Earth orbit, these prototypes reveal the subtle dynamics of compounding cycles in a weightless environment, laying the groundwork for deeper space deployment.

In the earliest terrestrial stages, prototypes are tested within vacuum chambers, simulating the space environment with cryogenic cooling to mimic the extreme cold of the cosmos. Here, engineers monitor the resilience of superconductors, assessing their stability and performance under continuous operation. These controlled experiments enable researchers to gather crucial insights on superconductive behavior in sustained cycles, leading to refinements in materials and configurations. For instance, certain superconductive alloys, initially found to degrade over extended cycles, prompt the development of more robust composites, each designed to maintain superconductivity with minimal resistance loss. These trials illustrate how every component—from the smallest conductor to the generator coils—must be meticulously tailored to support an uninterrupted flow of energy, ensuring that each cycle builds upon the last with precision.

Upon achieving promising results in simulated environments, these prototypes undergo testing in low-Earth orbit, where the absence of atmospheric interference and gravitational pull offers a clearer view of their true potential. This orbital testing unveils an unexpected efficiency boost as the motor-generator systems operate under minimal resistance, affirming the theoretical predictions about compounding in zero-gravity. Without terrestrial drag, each energy cycle completes with greater fidelity, allowing the system to compound at an unprecedented rate.

Observing this phenomenon in orbit not only reinforces the viability of energy compounding but reveals the exponential scalability inherent in the design. Each cycle yields a surplus that reinvests, doubling down on its initial promise and pushing the boundaries of energy generation as we know it.

Data collection from these orbital prototypes is critical. Autonomous sensors embedded throughout the system gather real-time metrics on temperature, energy flow, and resistance. This continuous stream of data feeds into an advanced AI-driven management system that regulates the prototype's operations, adjusting parameters to maintain peak efficiency. This adaptive intelligence becomes particularly essential when prototypes encounter fluctuations, such as minor solar radiation shifts or micro-meteoroid encounters, which could disrupt the delicate compounding cycle. Through machine learning algorithms, the system calibrates in real-time, preserving the continuity of energy flow and minimizing loss.

As testing progresses, scalability becomes a focal point. Engineers incrementally expand the prototypes, introducing modular components that enhance energy output while maintaining structural stability. These expansions allow for the inclusion of larger generators and more sophisticated transmission arrays, proving that energy compounding scales proportionally with system size. The compounding effect, now amplified by increased input, demonstrates the system's resilience and scalability, establishing a foundation for large-scale deployment that could one day meet global energy demands.

A breakthrough comes with the addition of wireless energy transmission capabilities, allowing the surplus energy generated in orbit to be beamed across short distances in space. Leveraging advancements in resonant inductive coupling, these trials confirm that energy can be transmitted without physical connectors, marking a transformative step toward the possibility of space-to-Earth energy beaming. Each successful test strengthens the potential for a system that not only generates sustainable energy in space but transmits it across vast distances, bridging the cosmic and the terrestrial with a seamless flow of power.

This stage of development embodies a new era in energy technology, one that combines the elegance of zero-gravity physics with the precision of superconductive engineering. By proving that energy compounding can function as an autonomous, scalable system, these prototypes establish the practical feasibility of an infrastructure that promises to redefine human access to energy. As each cycle compounds upon the last, a vision of boundless power emerges—a vision that beckons humanity to venture beyond the limitations of Earth-bound resources and into a realm where energy, once thought of as finite, becomes a limitless catalyst for progress.

The construction of a zero-gravity infrastructure capable of sustaining and scaling energy compounding demands meticulous planning, engineering precision, and an intricate understanding of both material science and the unique physics of space. Each structural element, from the superconductive coils to the transmission arrays, must withstand not only the rigorous conditions of deep space but also the demands of continuous, compounding energy cycles. This infrastructure—envisioned as a network of autonomous yet interconnected modules—represents a leap beyond conventional engineering, marrying resilience with flexibility to enable an enduring and expandable foundation for sustainable power.

Material selection plays a pivotal role in ensuring the longevity and reliability of this infrastructure. Superconductors, essential to minimizing resistance, are crafted from high-purity alloys and advanced ceramics specifically designed to retain stability under ultra-low temperatures. These superconductors operate with near-zero resistance, allowing energy to flow freely without dissipation. To support this delicate balance, structural materials must be equally robust, able to withstand the cosmic radiation, temperature extremes, and micro-meteoroid impacts characteristic of space. Carbon composites and reinforced polymers—lightweight yet exceptionally durable—provide the necessary resilience for frame components, while radiation-resistant coatings protect against gradual degradation, preserving the integrity of the infrastructure over extended missions.

The architecture of this infrastructure reflects a modular design philosophy, allowing for adaptability and scalability. Each module functions as an independent unit within the larger system, enabling engineers to replace, upgrade, or repair specific components without disrupting the overall energy cycle. This modularity, combined with a strategy of incremental assembly in space, optimizes both logistics and functionality. By deploying modules in stages, transported to stable orbits and assembled in situ by autonomous robots, the infrastructure becomes an evolving system that can expand its capacity in response to rising energy demands. This modular approach not only mitigates initial launch costs but also allows for phased integration of advanced technologies, aligning with a vision of perpetual growth and sustainability.

Strategic positioning of these modules at gravitational equilibrium points, such as Lagrange points, further enhances stability and efficiency. These points, where the gravitational forces between Earth and other celestial bodies balance, provide a natural anchor for orbiting stations, reducing the need for constant adjustments. Stationed at these equilibrium points, the infrastructure remains relatively fixed in position, benefiting from consistent exposure to sunlight and cosmic energy sources. This stability is crucial for maintaining a continuous energy cycle, as it minimizes the operational energy typically expended on station-keeping maneuvers, allowing more power to flow into the compounding process.

Embedded throughout this infrastructure are autonomous monitoring systems that harness artificial intelligence to optimize functionality. Sensors positioned within each module collect real-time data on everything from energy output to environmental conditions, feeding this information into a central AI-driven control system. This AI not only monitors system health but also makes dynamic adjustments in response to fluctuations, maintaining an optimal energy balance. In space, where human intervention is limited, this self-regulating capability is vital, enabling the infrastructure to operate autonomously, resiliently, and efficiently even in the face of variable cosmic forces. The AI's capacity for predictive maintenance also ensures that any potential issues are addressed before they escalate, preserving the integrity and continuity of the energy compounding cycle.

The infrastructure's design also considers the need for robust energy storage and transmission. Superconductive magnetic storage systems, integrated within the modules, allow surplus energy to be retained without dissipative loss, creating a buffer that enhances system stability. Transmission lines crafted from ultra-pure metals, oriented for minimal energy loss, facilitate the seamless transfer of power between modules, ensuring that each component contributes to the compounding effect. These superconductive lines, free from the limitations of gravity-bound conduits, enable a spatial configuration optimized for maximum energy flow, reinforcing the infrastructure's ability to operate as a cohesive, high-efficiency system.

In its entirety, this zero-gravity infrastructure stands as a testament to the potential of human ingenuity when liberated from terrestrial constraints. It offers a glimpse into a future where energy is no longer bound by the limitations of Earth's finite resources, instead deriving power from an environment where friction, gravity, and atmospheric interference no longer inhibit potential. This infrastructure not only facilitates the energy-compounding process but embodies the larger vision of an interconnected, space-based energy system capable of evolving and expanding in parallel with humanity's ambitions. Through this sophisticated fusion of material science, modular engineering, and autonomous regulation, the infrastructure marks the dawn of an era where energy abundance can become a tangible reality, reshaping the world below and extending humanity's reach into the cosmos.

The compounding energy infrastructure's journey from experimental models to fully realized zero-gravity stations demands rigorous adherence to fundamental physical laws, ensuring that each stage of design aligns with the immutable forces governing matter and energy. The laws of thermodynamics, electromagnetism, and quantum mechanics become essential guidelines, harmonizing this complex system with the natural universe. By integrating these principles at every level, the energy-compounding infrastructure achieves an unparalleled equilibrium between theoretical precision and practical functionality, forming a closed-loop system that not only operates efficiently but thrives on self-sustaining cycles.

The conservation of energy, a cornerstone of thermodynamics, underpins the entire compounding system. In a closed environment such as a zero-gravity energy station, energy is neither created nor destroyed but is instead meticulously transformed to maximize output while minimizing losses. This principle manifests through the seamless conversion of mechanical energy into electrical energy within the generator, achieving near-total efficiency by eliminating the energy losses commonly associated with terrestrial resistance and gravitational pull. The zero-gravity environment, therefore, enables a model of energy generation that echoes the self-sustaining loops found in nature, where each resource is perpetually recycled, sustaining itself in a cycle of regeneration.

Electromagnetic principles further refine the infrastructure, particularly through the use of electromagnetic induction, which lies at the core of energy transfer between the system's motor and generator. Superconductive materials, operating at ultra-low resistance levels, harness this induction process with extraordinary efficiency, leveraging Faraday's Law to convert mechanical motion into electrical energy with minimal loss. This careful calibration of electromagnetic interactions ensures that each energy cycle is both robust and sustainable, adhering to the fundamental requirements for continuous operation over extended periods. By fully exploiting the zero-resistance properties of superconductors in space, the system avoids the inefficiencies and dissipation that traditionally hinder power generation on Earth, achieving a compounding effect that grows with each cycle.

In the realm of motion and stability, Newtonian mechanics govern the behavior of the infrastructure, dictating the balance of forces essential for maintaining equilibrium in space. The infrastructure's rotation, orientation, and centripetal forces are all meticulously calibrated to achieve a state of balance that negates the need for constant corrections. By leveraging these forces strategically, the infrastructure stabilizes its position, allowing the energy cycle to proceed uninterrupted. This reliance on Newtonian principles not only enhances efficiency but also reduces operational demands, reinforcing the system's capacity to sustain itself without exhaustive resource expenditure.

Quantum mechanical insights, particularly those related to superconductivity, play an equally pivotal role. At the core of each superconductive component is the phenomenon of Cooper pairing, a quantum state that enables electrons to move through the material without encountering resistance. This property allows for the unimpeded flow of electricity within the infrastructure, supporting a compounding cycle that channels each energy surplus back into the system. By harnessing the zero-resistance state made possible by quantum mechanics, the infrastructure achieves an operational fidelity that traditional materials cannot replicate, thus pushing the boundaries of energy efficiency.

Entropy, a concept deeply rooted in thermodynamics, also guides the design. In terrestrial systems, entropy leads to inevitable degradation, as energy dissipates through heat and other forms of loss. However, the zero-gravity infrastructure mitigates entropy by creating a system where energy is meticulously cycled with minimal waste, optimizing thermal management to retain operational stability. By maintaining a lower entropy state, the infrastructure extends its functional lifespan, ensuring that each cycle compounds without succumbing to the usual degradations seen in Earth-bound systems.

In addition, the infrastructure's alignment with magnetic resonance and coherence creates a finely tuned harmony in the oscillations and frequencies of the compounding process. By syncing with the natural oscillatory patterns inherent in its components, the system achieves a state of resonance that enhances energy transfer while minimizing mismatches and losses. This coherence, achieved through precise calibration of the generator and motor cycles, reinforces the compounding effect, as each cycle builds seamlessly upon the last.

This rigorous adherence to physical laws establishes a foundation that transcends mere functionality, rendering the infrastructure an embodiment of scientific elegance and technological foresight. Through its alignment with universal principles, the zero-gravity energy-compounding infrastructure not only achieves unparalleled efficiency but also reflects a model of sustainable operation that harmonizes with the natural order. Each component, cycle, and interaction within the system converges toward a singular objective: to transform energy from a finite resource into a boundless force for human advancement.

The journey from theoretical model to functional infrastructure has yielded early prototypes, each one a testament to the gradual refinement of the zero-gravity energy-compounding concept. These prototypes, initially tested within terrestrial labs and then in low-Earth orbit, offer invaluable insights into the dynamics of energy generation, transfer, and compounding in a near-perfect vacuum. Each iteration in this prototyping process serves to crystallize the principles underlying the compounding model, proving that a self-sustaining energy cycle can indeed thrive when liberated from Earth's constraints.

Within controlled laboratory settings, engineers replicate the low-gravity, low-temperature conditions of space, utilizing vacuum chambers and cryogenic cooling to simulate the environment that awaits the infrastructure in orbit. In these meticulously engineered environments, superconductive coils undergo rigorous testing, allowing scientists to observe their behavior under continuous operation. The data collected from these trials reveal the effects of sustained energy flow on superconductive materials, leading to critical adjustments in their composition and structure. These refinements ensure that the superconductors maintain their stability across cycles, preserving their ability to carry high currents without resistance—a fundamental requirement for achieving compounding efficiency.

With promising results on Earth, the prototypes advance to testing in low-Earth orbit, where the absence of atmospheric drag and gravitational interference offers a closer approximation to deep-space conditions. Here, the motor-generator systems, operating free from terrestrial resistance, demonstrate a remarkable efficiency boost, with each cycle producing a measurable surplus. This orbital phase not only validates theoretical predictions but also unveils the potential scalability of energy compounding, as the surplus generated in each cycle reinvests itself, enabling the system to grow exponentially over time. In space, where friction is nearly nonexistent, the prototypes' energy cycles achieve a fluidity and coherence that terrestrial systems can rarely match, underscoring the viability of this infrastructure as a self-sustaining entity.

Autonomous sensors embedded within the prototypes capture real-time data on temperature, energy output, and material stability, feeding this information into a central AI-driven monitoring system. The AI, trained to recognize even the subtlest deviations in system performance, makes instant adjustments to optimize efficiency, effectively mitigating fluctuations in energy flow. This adaptability is crucial in orbit, where exposure to cosmic radiation and micro-meteoroid impacts could otherwise disrupt operations. By responding dynamically to these variables, the AI preserves the integrity of the energy cycle, ensuring that each prototype remains stable and functional over time.

As the testing phase progresses, scalability becomes a focal point. Engineers expand the prototypes with modular components designed to enhance energy output without sacrificing efficiency. This scalability is carefully monitored, as each increase in input tests the infrastructure's capacity to compound energy on a larger scale. These trials reveal a crucial insight: the compounding effect not only holds but actually amplifies as the system grows. The scalability of the infrastructure thus becomes its defining feature, presenting a clear path toward meeting the global energy demands of the future.

A key milestone in prototype testing involves the integration of wireless energy transmission capabilities. Building on the theoretical foundation laid by Nikola Tesla, these prototypes demonstrate the feasibility of transmitting energy over distances without the need for physical connectors. Using resonant inductive coupling, the prototypes successfully beam energy across short distances in orbit, paving the way for space-to-Earth transmission. This breakthrough, although in its nascent stages, marks a transformative step toward an infrastructure that can generate power in space and deliver it to receivers on Earth, eliminating the need for resource-intensive terrestrial grids.

Each prototype, in its success and evolution, embodies the promise of energy abundance. By refining the principles of compounding, scaling, and wireless transmission, these models lay the groundwork for a full-scale infrastructure capable of revolutionizing humanity's relationship with power. With each successful test, the vision of a self-sustaining, universally accessible energy system moves closer to reality, challenging the scarcity-driven paradigms that have long defined human progress.

For the energy compounding infrastructure to transition from experimental prototype to full-scale deployment, it must not only meet technical and scientific benchmarks but also navigate a complex landscape of international collaboration and resource allocation. The scope and scale of a space-based energy network transcend national borders, presenting a unique opportunity for nations to unite under a shared vision of sustainable power. Establishing such an infrastructure calls for unprecedented cooperation, where knowledge, resources, and shared responsibility converge to create a system capable of meeting global energy demands and catalyzing collective progress.

The necessity of global partnerships is foundational to the success of this ambitious endeavor. No single nation possesses the resources, expertise, or infrastructure to establish an autonomous energy-compounding system in space. The scale of this project requires a consortium of public and private entities, each contributing specialized knowledge, technological assets, and financial support. By fostering collaborative alliances, the infrastructure gains the strength of diverse expertise, with each participating entity reinforcing the project's viability and resilience. Through this unified approach, the vision of energy abundance can be realized as a shared asset, transcending political boundaries to benefit all of humanity.

Funding and resource allocation present both challenges and opportunities. The infrastructure's development necessitates substantial investment in research, materials, and launch capabilities, as well as the establishment of ground-based receiving stations for wireless energy transmission. By pooling resources, countries can mitigate individual financial burdens while maximizing the infrastructure's potential impact. International funding models—perhaps guided by the precedent set by large-scale projects such as the International Space Station—can distribute costs equitably, allowing both developed and developing nations to participate. This collaborative financing structure enables a more equitable distribution of energy, ensuring that the benefits of this infrastructure reach all corners of the globe.

Coordinating multinational efforts also requires carefully negotiated governance frameworks. Such a system, once operational, has the potential to disrupt existing energy markets and geopolitical dynamics, making it essential to establish clear policies for managing, regulating, and sharing the output. International treaties could govern aspects of the infrastructure, ensuring that no single nation or corporation monopolizes access or control. These treaties would enshrine principles of fairness and transparency, providing a blueprint for ethical stewardship that aligns with the infrastructure's vision of universal access and sustainability.

Shared responsibility and ownership, another critical aspect of collaboration, reinforce the infrastructure's long-term resilience. By designating joint accountability for its maintenance, security, and operations, participating nations commit to the infrastructure's stability and integrity. This collective ownership model not only ensures that all stakeholders have a vested interest in the infrastructure's success but also fosters a spirit of global citizenship, where energy abundance is seen as a common good rather than a commodity to be monopolized. Such an approach aligns with the ethical imperatives of the infrastructure, creating a system that not only addresses humanity's immediate energy needs but also establishes a sustainable foundation for future generations.

Overcoming political and logistical barriers will be vital in turning this vision into reality. Political tensions, regulatory differences, and logistical challenges pose significant obstacles to multinational collaboration, yet these challenges are not insurmountable. Through diplomatic channels, technology-sharing agreements, and joint research initiatives, nations can work together to navigate these complexities, forging pathways for cooperation that transcend individual interests. By embracing a model of collaborative governance, the infrastructure can emerge as a unifying force, demonstrating how energy abundance can inspire cooperation and shared responsibility on a global scale.

In the end, the realization of a zero-gravity energy-compounding infrastructure is more than a technological feat; it is a testament to humanity's capacity for unity in the pursuit of shared ideals. This vision invites nations to transcend historical divisions, channeling collective efforts toward a common goal of sustainable abundance. As nations come together to build this infrastructure, they participate in a legacy of collaboration that stands as a beacon for future generations—a blueprint for how innovation, when guided by mutual respect and shared purpose, can transform not only technology but the very nature of human progress.

The implementation of a zero-gravity energy-compounding infrastructure extends beyond scientific and technical innovation; it demands a profound commitment to ethical responsibility and stewardship. As humanity steps into an era of potential energy abundance, there lies an imperative to govern this infrastructure in a way that honors its transformative potential without compromising its integrity. This responsibility necessitates a framework that safeguards against misuse, ensures equitable access, and aligns with the foundational principles of sustainability, equity, and global welfare.

The ethics of energy abundance hinge on the idea that access to power should be a universal right, not a privilege. The infrastructure's compounding effect, if fully realized, holds the power to redefine the distribution of resources on a global scale, offering clean, accessible energy to even the most remote and underserved regions. This paradigm shift could bridge socioeconomic divides that have long separated communities and countries, empowering populations with the resources needed to pursue health, education, and economic development. However, to fulfill this vision, governing entities must establish policies that protect against monopolization, ensuring that no single entity or group of entities can claim exclusive rights to this resource. By codifying energy abundance as a shared human asset, such policies would prevent exploitation and create a foundation for a more just and equitable world.

Central to this ethical framework is the principle of transparency. In the absence of transparency, even the most well-intentioned systems risk falling prey to misuse, eroding public trust and jeopardizing the infrastructure's long-term viability. By adopting open governance practices that allow for public oversight, participating nations can create a structure that is accountable to the world's citizens. Through clear, accessible reporting on energy output, distribution, and governance decisions, this infrastructure can serve as a model for ethical management, reinforcing its commitment to the ideals of fairness and inclusivity.

Safeguarding against exploitation also involves proactive measures to prevent energy from becoming a tool of geopolitical control. History is replete with instances where essential resources have been wielded as instruments of power, exacerbating tensions and conflict. The infrastructure's design must account for these historical precedents, incorporating safeguards that ensure energy cannot be weaponized or used as leverage in geopolitical disputes. To this end, international treaties can serve as legal protections, setting boundaries on how energy generated by the infrastructure may be utilized. By positioning energy abundance as a neutral, peace-promoting asset, humanity can transcend the patterns of the past, embracing a new era where resources unite rather than divide.

In addition to these protections, an ethical framework must prioritize environmental stewardship. The infrastructure's capacity to generate limitless clean energy presents an opportunity not only to power human civilization but to restore balance to the planet's ecosystems. With abundant energy, humanity can support large-scale environmental restoration projects, from reforestation to carbon capture, counteracting decades of environmental degradation. This infrastructure thus becomes not only a solution to energy scarcity but a catalyst for environmental rejuvenation, aligning human progress with the natural world rather than opposing it.

Moreover, responsible innovation is essential to maintaining the infrastructure's sustainability. As technology evolves, so too must the protocols governing its use, ensuring that advancements in energy generation, transmission, and storage are applied ethically and equitably. By establishing adaptive governance structures that evolve with technological progress, the infrastructure can remain resilient and relevant, fostering a legacy of responsible development. This model of innovation, grounded in ethical foresight, ensures that energy abundance will continue to serve the common good, benefiting humanity without compromising future generations.

Ultimately, the ethical dimensions of zero-gravity energy compounding invite humanity to engage with technology not merely as a tool but as a transformative force for positive change. By adhering to principles of transparency, inclusivity, and environmental responsibility, this infrastructure can fulfill its potential as a beacon of ethical progress. In a world no longer constrained by energy scarcity, the infrastructure offers a vision of possibility—one where human ambition aligns harmoniously with the planet's well-being and the ideals of global equity.

The vision of a zero-gravity energy-compounding infrastructure extends beyond immediate technological advancements, hinting at a transformative legacy that reshapes civilization's trajectory. With energy abundance realized, society stands on the precipice of an unprecedented era, where humanity's progress is no longer limited by the availability of resources but is instead driven by the sustainable, regenerative force of compounded power. This legacy, grounded in both ethical stewardship and scientific rigor, opens the door to a future defined by resilience, equity, and boundless opportunity.

The impact of energy abundance on global industries could be profound. Sectors traditionally constrained by high energy costs—such as agriculture, manufacturing, and transportation— would find new freedoms in a world where energy is not a limitation but a catalyst. Imagine agriculture powered by limitless energy, where vertical farms provide food security in urban centers, and desalination plants offer clean water to arid regions. The manufacturing industry could adopt energy-intensive processes without environmental compromise, producing goods sustainably and at scale. Transportation, too, could undergo a revolution, with clean energy propelling vehicles, ships, and planes without the need for fossil fuels, drastically reducing carbon emissions and reshaping global logistics.

Beyond industry, the infrastructure's reach could extend into daily life, improving quality of life and supporting innovations that uplift society's most vulnerable. Communities that once struggled with unreliable access to electricity would find themselves empowered, with energy abundance creating avenues for education, healthcare, and economic growth. Reliable access to power could transform remote regions into hubs of opportunity, where digital learning, telemedicine, and technological infrastructure bridge gaps that geography once made insurmountable. As society reaps these benefits, the infrastructure's true legacy emerges: a world where human potential is no longer tethered by resource scarcity but inspired by the possibilities of sustainable abundance.

The infrastructure also redefines humanity's relationship with the environment. In a model where energy flows abundantly and cleanly, society can invest in large-scale environmental restoration efforts, turning technology's gaze toward the planet itself. This vision includes expansive reforestation projects, carbon capture technologies, and even efforts to restore biodiversity in ecosystems strained by industrialization. With energy as a regenerative force, society could reclaim degraded lands, clean polluted waterways, and reinvigorate habitats that have been diminished by human activity. This model of stewardship aligns human advancement with ecological healing, positioning energy abundance as both a solution to environmental crises and a foundation for harmonious coexistence.

Yet, perhaps the most profound impact of energy abundance lies in its influence on human potential. Freed from the demands of survival, individuals and communities could turn their focus toward pursuits that enrich society intellectually, culturally, and spiritually. The infrastructure's promise is not merely to fuel industry or power cities but to enable a paradigm shift, where resources are channeled toward the arts, sciences, and explorations that expand the human experience. Education, research, and creative expression would flourish, as energy abundance grants individuals the space and resources to innovate, explore, and dream. This is a legacy that transcends generations—a testament to what humanity can achieve when it no longer views energy as a constraint but as an enabler of boundless growth.

As humanity builds this infrastructure, it embarks on a journey that reflects the highest aspirations of technological advancement tempered by ethical vision. The zero-gravity energy-compounding system, in its full realization, becomes more than a technological marvel; it becomes a beacon of possibility, illustrating how civilization can rise above the limitations of scarcity and embrace a model of sustainable, inclusive prosperity. This infrastructure serves as a call to reimagine what is possible, inspiring a world where technology and nature are no longer at odds, but are partners in a shared future of renewal and growth.

In a world transformed by energy abundance, the zero-gravity energy-compounding infrastructure stands as a blueprint for a sustainable future, offering more than mere power; it provides a framework for a civilization grounded in resilience and innovation. The infrastructure's design, centered on compounding principles, reveals an inherent adaptability, positioning it not only as a solution for today's energy challenges but as a foundation for humanity's long-term stability. This infrastructure invites future generations to expand, innovate, and build upon it, perpetuating a cycle of development that respects both human needs and planetary boundaries.

One of the most compelling aspects of this legacy is the infrastructure's ability to foster a post-scarcity mindset. In breaking free from the constraints of finite resources, society can shift its focus from competition over limited assets to collaboration in expanding shared potential. This shift echoes throughout economic, social, and environmental realms, redefining what it means to prosper in harmony with others. In this model, nations no longer strive to hoard energy resources or dominate energy markets; instead, they invest in cooperative growth, recognizing that in an interconnected, abundant energy network, collective benefit supersedes isolated gain.

The infrastructure's influence on governance could reshape international relations, introducing a model of cooperative energy stewardship. By establishing shared governance structures, nations create a foundation for peace, stability, and mutual respect, where power is not wielded as leverage but offered as a universal right. This cooperative framework challenges existing geopolitical paradigms, demonstrating that technology, when managed with ethical foresight, can become a catalyst for unity rather than division. The infrastructure's governance structures thus become a model for addressing other global issues, fostering a spirit of collaboration that may extend beyond energy to encompass shared approaches to climate change, resource distribution, and planetary stewardship.

The journey toward this energy-compounding infrastructure exemplifies the human capacity for vision, persistence, and ethical responsibility. Each phase, from theoretical conception to physical implementation, reflects a commitment to creating technology that serves humanity without compromising future generations. As society begins to harness this infrastructure's potential, it does so with an understanding that its legacy will extend far beyond the boundaries of energy. The zero-gravity system becomes an icon of sustainable progress, symbolizing a paradigm shift where technology no longer exploits but empowers, where human ambition aligns with environmental responsibility, and where prosperity is measured not by the accumulation of resources but by the potential unlocked within every individual.

The dawn of energy abundance invites humanity to consider a profound question: How will we define our future in a world no longer bound by scarcity? The infrastructure's legacy offers a powerful answer. By embracing the principles of energy compounding, humanity can transcend traditional boundaries, pursuing endeavors that were once constrained by resources and redefining what it means to progress as a global society. In this world, energy is more than a utility; it is a force that powers human potential, illuminates paths to equality, and fosters a civilization where creativity, learning, and exploration are open to all.

As the zero-gravity energy-compounding infrastructure reaches its full potential, it leaves a legacy that speaks to the heights humanity can achieve when guided by ethical foresight and collaborative spirit. The infrastructure becomes more than a source of power; it is a testament to human ingenuity and a promise of a future where resources empower rather than limit. This vision of abundance, sustained by the infrastructure's enduring cycles, challenges society to pursue a new era of stewardship, where technology is harnessed not merely for survival but to elevate the collective human experience.

Chapter 7: Vision for the Future— A Blueprint for Global Transformation

The vision of a future defined by global energy abundance is not just a conceptual leap; it is a call to action, a roadmap toward transforming humanity's reliance on finite resources into a sustainable network of power accessible to all. This transformation demands an infrastructure that reaches beyond the constraints of Earth and into the boundless expanse of space, where energy can be generated, compounded, and delivered without the limitations that have shaped humanity's energy systems for millennia. To realize this future, a meticulously designed pathway for deploying, expanding, and governing this infrastructure is essential—each phase of implementation must be measured, each benchmark carefully aligned with the broader vision of universal energy access and environmental harmony.

Creating an infrastructure capable of delivering energy abundance on a planetary scale begins with small, precise steps that lay the foundation for larger expansions. Initial testing phases are conducted in controlled environments where components like superconductive coils and motor-generators, optimized for zero-gravity conditions, undergo rigorous trials. Here, engineers simulate the extreme conditions of space, replicating the low temperatures, vacuums, and gravitational neutrality necessary for validating each component's functionality. Through these trials, each system element is refined to withstand the demands of continuous, frictionless operation, ensuring reliability and efficiency before larger investments are committed.

Once initial tests confirm component durability and efficiency, the infrastructure advances to a pilot phase that deploys prototype systems directly in space. These initial modules, engineered for modularity, are assembled incrementally, forming a cohesive unit that can be monitored for energy output, stability, and transmission efficiency. This stage marks the first instance of energy beaming from space to Earth, a proof of concept that validates the system's capability to deliver power across distances. Relay nodes positioned at strategic points support this energy flow, ensuring that power reaches Earth-based receivers consistently, with the flexibility to adjust for real-time variables and interruptions. These early successes create benchmarks that guide the expansion of the infrastructure, confirming that it can operate not only reliably but sustainably.

As confidence in the infrastructure's capacity grows, the deployment enters a phase of systematic scaling, where new modules are integrated according to a carefully orchestrated schedule. In this phased expansion, each additional module is equipped with advanced monitoring systems that ensure it aligns seamlessly with the existing network. This approach minimizes disruptions while supporting the infrastructure's ability to grow in sync with Earth's increasing energy demands. Simultaneously, terrestrial receivers are expanded to cover a wider geographical range, creating a robust interface that allows the space-based infrastructure to power larger sections of the global grid without interruption.

For long-term operational success, these systems are designed to function autonomously, with minimal human intervention. Autonomous diagnostic and maintenance technologies, such as repair drones and self-monitoring sensors, extend the infrastructure's durability and reduce dependence on costly, logistically complex repairs. These systems perform routine assessments, detecting and addressing potential issues to keep each component at optimal performance. Such advancements are critical to maintaining the infrastructure's resilience, allowing it to serve as a reliable energy source without constant oversight. Through automation and continuous diagnostics, the infrastructure evolves into a self-sustaining system capable of enduring over generations.

International collaboration underpins each phase of deployment, with countries contributing both financial resources and technical expertise. By establishing a shared ownership model, participating nations solidify the infrastructure as a global asset, with benefits that extend beyond national borders. This collaborative framework encourages transparency, accountability, and equitable distribution, ensuring that the infrastructure's resources support global progress rather than individual interests. Through this partnership, humanity steps toward a future where energy is universally accessible, creating a foundation for a world defined by shared prosperity and sustainable growth.

Achieving large-scale deployment of space-based energy infrastructure requires more than technical precision; it demands clear and measurable benchmarks to assess the project's trajectory. These benchmarks serve as markers of progress, establishing targets for energy output, transmission reliability, and the seamless integration with Earth's diverse power grids. By setting and maintaining these benchmarks, engineers, scientists, and policymakers ensure that each phase of development advances systematically, supporting the vision of sustained energy abundance for all.

The initial benchmarks are focused on energy yield in zero-gravity conditions. In space, where superconductive materials encounter minimal resistance, the infrastructure is designed to operate with near-total efficiency, with energy cycles yielding a compounded surplus. Early benchmarks monitor this yield, validating that each cycle produces energy that exceeds consumption, reinforcing the infrastructure's compounding capacity. By meeting these targets, the infrastructure achieves the fundamental prerequisite for sustainable operation: a continuous and reliable energy surplus.

Following initial energy yield tests, benchmarks shift toward operational stability and consistency, particularly under extended use. In space, the infrastructure must demonstrate an ability to withstand environmental variables, from temperature fluctuations to solar radiation. Stability benchmarks are achieved by subjecting each component to prolonged testing, simulating the demanding conditions it will encounter over years of uninterrupted operation. This focus on durability ensures that the infrastructure can deliver a steady supply of energy over time, preventing disruptions that could undermine the continuity of global energy distribution.

The successful integration with terrestrial grids forms the next set of benchmarks. As the infrastructure begins transmitting energy to Earth, these benchmarks measure the system's compatibility with a variety of grid standards and capacities worldwide. Transmission benchmarks focus on achieving efficient, stable energy transfer, while also addressing regional requirements to accommodate differences in grid infrastructure. Achieving seamless integration with terrestrial grids is crucial, as it ensures that the energy produced in space can be reliably distributed to meet Earth's diverse energy needs.

In the final phases, long-term benchmarks expand to include environmental impact and the scalability of the infrastructure. These benchmarks address the infrastructure's capacity to scale responsibly, without causing adverse ecological effects. Energy transmission to Earth, for example, must account for atmospheric interference and other potential environmental impacts, ensuring that the infrastructure operates harmoniously with natural systems. These safeguards prevent unintended consequences, aligning the infrastructure with principles of environmental responsibility and ensuring that it remains a force for sustainable development.

Through this framework of carefully calibrated benchmarks, the infrastructure evolves from a nascent idea into a robust, scalable system. By rigorously measuring progress, each phase of development contributes to the infrastructure's reliability, scalability, and environmental compatibility, establishing a stable foundation for an era defined by energy abundance.

Creating a global network for energy distribution is integral to realizing the full potential of space-based power. This network, spanning from orbit to remote areas of Earth, offers more than just a conduit for energy; it establishes a resilient infrastructure capable of delivering abundant, clean power across continents, reaching regions long hindered by resource scarcity. By integrating the space-based infrastructure with terrestrial power grids, this network introduces a balanced, adaptive system that can respond to fluctuating demand and deliver energy where it is needed most, making it a cornerstone of a sustainable global future.

The architecture of this global network prioritizes adaptability and responsiveness, essential for aligning energy supply with the diverse needs of Earth's regions. At its core, the infrastructure is supported by orbital relay nodes that serve as hubs for energy transmission. Positioned in geostationary orbit, these nodes receive power from space-based generation stations and direct it to designated terrestrial receivers. This positioning provides uninterrupted access to solar energy, while the nodes' advanced relay capabilities ensure that energy flows continuously, without delays or interruptions. The orbital hubs form the backbone of the network, facilitating a stable energy flow that can adjust in real-time to demand fluctuations.

This network's design emphasizes efficient, seamless integration with terrestrial grids, an undertaking that requires cooperation between nations, regulatory bodies, and power companies. To accommodate regional variations in energy infrastructure, international standards for transmission protocols are established, creating a cohesive framework that enables compatibility across diverse grids. These standards foster interoperability, allowing each part of the network to function harmoniously, regardless of local grid configurations. By adopting such a unified approach, the network not only maximizes energy distribution efficiency but also ensures that no region is excluded from access to sustainable power.

As the global network matures, it provides a platform for equitable energy distribution that addresses historical inequalities. Through widespread access to power, communities previously disconnected from reliable energy resources gain the opportunity to pursue economic development and improve living standards. Rural and remote areas that once lacked consistent energy access are now able to establish and sustain essential services such as healthcare, education, and clean water supply. By transforming access to power, the network becomes a catalyst for social equity, bridging gaps that once separated communities and enabling a more inclusive global society.

Beyond its impact on energy distribution, the global network also promotes environmental sustainability. By facilitating the widespread transition from fossil fuels to renewable energy, this infrastructure plays a direct role in reducing greenhouse gas emissions and alleviating environmental degradation. Clean energy from space empowers regions to minimize their reliance on nonrenewable sources, supporting large-scale efforts to reduce the human impact on the planet's ecosystems. Furthermore, the infrastructure's adaptable design allows it to evolve in response to future environmental needs, positioning it as a model of ecological resilience.

In this way, the global energy network becomes more than an infrastructure project; it becomes a testament to humanity's ability to collaborate across borders, prioritize sustainability, and foster a future where energy serves as a vehicle for progress and unity. Through this network, space-based energy reaches beyond technical achievement, shaping a new era of shared prosperity and environmental stewardship that transcends generations.

Establishing a sustainable model for the expansion of space-based energy infrastructure is essential for maintaining a trajectory of growth that aligns with environmental, economic, and social considerations. This approach prioritizes a phased, modular expansion that responds to increasing demand without imposing undue strain on planetary or financial resources. By embracing modularity, this infrastructure can evolve organically, adapting to regional needs and advancing in a way that safeguards its foundational principles of sustainability, accessibility, and ecological balance.

The infrastructure's modular design is central to its sustainable expansion. Each module, equipped with energy generation, storage, and transmission capabilities, functions as a self-contained unit. This modularity allows the infrastructure to grow incrementally, scaling up as demand rises and adapting to specific geographical requirements without necessitating the overhaul of existing systems. Modules can be added or upgraded based on location-specific energy needs, ensuring that expansion proceeds in a measured, efficient manner. This approach mitigates the risks associated with rapid, large-scale deployment, preserving system stability and resource allocation.

Resource management also plays a critical role in ensuring that expansion remains sustainable. The advanced materials used in constructing the infrastructure—such as superconductors and high-durability alloys—are chosen not only for their performance but for their longevity and recyclability. A circular approach to resource management governs the infrastructure's lifecycle, with components designed to be refurbished, recycled, or repurposed as they reach the end of their operational lifespan. By minimizing waste and encouraging reuse, this resource model aligns with the broader environmental goals of the infrastructure, supporting its growth without compromising ecological integrity.

Economic sustainability within this expansion model is equally vital. Funding for the infrastructure is distributed across a coalition of international partners, with contributions that reflect the project's global impact. Through a phased financial model, investments are synchronized with each stage of expansion, ensuring that financial resources align with project needs and avoiding the pitfalls of overextension. Revenue generated from energy distribution provides a reinvestment stream, reinforcing the infrastructure's capacity to finance further expansion independently. This self-sustaining financial model not only supports growth but enhances the infrastructure's long-term resilience.

Environmental considerations are woven into every aspect of the expansion model, ensuring that ecological balance remains a priority as the infrastructure grows. On Earth, terrestrial receivers are strategically located in regions that minimize habitat disruption, such as deserts or uninhabited lands, thus avoiding interference with ecosystems. Furthermore, the infrastructure's reliance on clean, renewable power significantly reduces carbon emissions, enabling regions to transition from fossil fuels and contribute to global efforts in combating climate change. Through these environmental protections, the infrastructure exemplifies a commitment to ecological harmony, illustrating how technological progress can coexist with respect for natural systems.

Equity in energy access remains a guiding principle as the infrastructure expands. As new modules are deployed, priority is given to regions that historically lacked reliable power, thereby promoting global energy equity. Agreements between participating nations ensure that energy benefits are distributed fairly, with an emphasis on supporting communities in need. This approach transforms energy access from a privilege into a universal right, aligning the expansion model with ideals of social justice and inclusion.

By integrating these elements—modularity, resource conservation, financial planning, environmental mindfulness, and equitable access—the expansion model lays a pathway toward a global energy system that serves both present and future generations. This sustainable model of growth sets a precedent for how large-scale technology can develop responsibly, establishing an infrastructure that not only meets the needs of a growing population but honors the principles of sustainability, resilience, and shared human progress.

The deployment of space-based energy infrastructure calls for a governance structure rooted in ethical oversight and transparency. As this transformative technology scales to serve a global audience, the principles guiding its administration become as essential as the technology itself. A framework for governance, built upon international collaboration, transparency, and ethical stewardship, ensures that this infrastructure will serve as a shared resource dedicated to the common good rather than becoming a tool of power or control for any single entity.

At the heart of this governance framework is the commitment to equitable access. Space-based energy has the potential to deliver clean, abundant power across borders, making it imperative that no single nation or corporation monopolizes its benefits. To safeguard this inclusivity, participating countries establish agreements that govern energy distribution according to population needs, development levels, and regional energy requirements. This allocation framework prevents any party from accumulating undue influence over the infrastructure, ensuring that energy abundance is available to all as a shared global asset.

Environmental protection is another cornerstone of this governance model. The infrastructure's impact on natural ecosystems, both on Earth and in orbit, must be carefully managed. Policies guide the sustainable construction and maintenance of space-based infrastructure, with protocols that minimize interference with Earth's atmosphere and protect sensitive terrestrial and aquatic habitats. These regulations are overseen by an international coalition responsible for conducting environmental assessments and ensuring compliance with established standards. This commitment to ecological stewardship reinforces the infrastructure's alignment with sustainable development goals, ensuring that humanity's pursuit of energy abundance respects and preserves the planet's natural resources.

Transparency in governance is achieved through open access to information about the infrastructure's operations. An independent oversight body, representing participating nations and stakeholders, regularly publishes detailed reports on energy production, distribution, and ecological impact. These reports are accessible to the public, fostering accountability and ensuring that the infrastructure's benefits are equitably shared. By maintaining transparency, the governance structure builds trust among nations and citizens, reinforcing the infrastructure's role as a resource dedicated to the welfare of all humanity.

To uphold ethical standards, strict guidelines define permissible applications of space-based energy, prioritizing projects that support human well-being, environmental restoration, and economic growth. By focusing on constructive and socially beneficial uses, the governance structure ensures that the energy generated by this infrastructure serves humanity's highest ideals. Mechanisms are in place to prevent misuse, with penalties for entities found diverting resources toward purposes that contravene these ethical standards. This vigilant oversight ensures that the infrastructure remains aligned with its intended mission as a force for positive global transformation.

Through a comprehensive governance model that emphasizes equity, environmental protection, transparency, and ethical responsibility, the space-based energy infrastructure sets a new standard for how global resources can be managed. This model provides a framework for addressing future technological advancements, guiding humanity toward a path where innovation serves the collective good. By establishing these governance principles, participating nations create a legacy of ethical stewardship, illustrating that with foresight and collaboration, humanity can harness its greatest advancements to uplift all people and protect the natural world.

Protecting the infrastructure from misuse and exploitation is essential to ensuring that its benefits remain focused on advancing global welfare. As this space-based energy system scales to unprecedented levels, so too does the responsibility to guard it against potential monopolization, misuse, and unethical applications. A framework of safeguards, grounded in transparency, rigorous monitoring, and ethical standards, protects the infrastructure from being co-opted by any single interest, thus preserving its integrity as a universal resource.

Central to these safeguards is a governance model that prevents any single nation or corporation from exerting undue control over energy resources. Through binding international agreements, this model establishes a coalition of equal stakeholders, each with defined roles and responsibilities. These agreements emphasize that the infrastructure serves humanity as a whole, forbidding actions that would redirect resources for exclusive or exploitative purposes. This shared authority ensures that no single entity can manipulate the infrastructure for unilateral gain, promoting an atmosphere of mutual benefit and respect among all contributors.

Real-time monitoring systems reinforce these safeguards by providing continuous oversight of energy production, allocation, and usage patterns. Using advanced algorithms and data analytics, these systems detect any deviations from intended use, such as unauthorized attempts to reroute power or store energy excessively. By enabling rapid response to irregularities, this monitoring framework ensures that all resources are allocated in accordance with established ethical guidelines. Furthermore, these systems offer a layer of transparency, as participating nations can access real-time data to verify that the infrastructure's operations align with international agreements and ethical principles.

Strict regulatory guidelines further strengthen the infrastructure's protection against misuse. These guidelines define acceptable applications for the generated energy, prioritizing uses that contribute to social welfare, economic development, and environmental stewardship. Regulations prohibit applications that could lead to harm or exploitation, with penalties for entities that attempt to circumvent these standards. By enforcing clear boundaries, the infrastructure's governance model ensures that the energy serves only constructive and responsible purposes, reinforcing its role as a resource for global good.

Environmental considerations are equally crucial in preventing misuse. Policies restrict energy transmission methods to those that minimize ecological disruption, with assessments conducted regularly to monitor environmental impact. These environmental safeguards are embedded into the infrastructure's operational protocols, requiring that adjustments be made if any transmission methods are found to interfere with natural ecosystems. This approach underscores a commitment to balancing technological advancement with ecological preservation, ensuring that the infrastructure's growth does not come at the expense of environmental health.

Through this combination of transparent governance, vigilant monitoring, ethical regulations, and environmental protections, the infrastructure achieves a robust defense against misuse. These safeguards ensure that as space-based energy becomes integral to human progress, it remains a source of empowerment and equity, free from the risks of exploitation or environmental harm. In this protected framework, the infrastructure becomes a model of ethical innovation, showcasing how humanity's most ambitious endeavors can advance not only technological horizons but also ethical standards that honor and protect the collective good.

As humanity embraces space-based energy infrastructure, establishing ethical standards and practices becomes paramount to guiding its use and expansion responsibly. The introduction of such transformative technology offers an opportunity to redefine the ethical landscape of energy generation and distribution, setting precedents that prioritize human welfare, environmental sustainability, and societal equity. By anchoring the infrastructure in a robust ethical framework, stakeholders ensure that it serves as a force for positive change, aligning technological advancement with the principles of fairness, accountability, and foresight.

Central to these ethical standards is the commitment to prioritize human welfare above all else. The infrastructure's energy resources are directed toward applications that support global development, health, and education, ensuring that its power fuels endeavors that uplift and enhance the human experience. Clear guidelines govern the allocation of energy, favoring projects that address urgent societal needs, from poverty alleviation and healthcare support to educational outreach and technological empowerment in underserved regions. This focus on human welfare transforms energy into a catalyst for societal growth, reinforcing the infrastructure's role as a foundation for global progress.

Environmental stewardship is an equally vital component of these ethical standards. Recognizing the infrastructure's potential impact on Earth's ecosystems, its operations are governed by protocols that minimize ecological footprint and promote sustainability. This commitment to environmental responsibility is reflected in the infrastructure's design, resource management, and operational practices, which all emphasize minimal interference with natural systems. Regular environmental assessments guide adjustments to ensure that the infrastructure's expansion harmonizes with ecological preservation, reinforcing the ethical obligation to protect the planet for future generations.

Transparency remains a guiding principle within this ethical framework. Open communication and public access to information about the infrastructure's operations, energy output, and environmental impact foster accountability and trust. An independent oversight body, composed of representatives from participating nations and organizations, regularly publishes findings and decisions, allowing stakeholders and the public to assess the infrastructure's adherence to its ethical commitments. This transparency not only builds confidence but also establishes a model of responsible governance that can inspire future initiatives.

The infrastructure's ethical standards extend to its role in addressing global inequality. By committing to equitable energy distribution, the infrastructure bridges historical divides that have left many communities without reliable power. Special provisions ensure that developing regions, rural areas, and economically disadvantaged populations benefit proportionately from the infrastructure's resources. This model of inclusivity turns energy into a vehicle for equality, creating pathways for prosperity and resilience in regions long marginalized by energy scarcity. In this way, the infrastructure serves as a beacon of ethical progress, showcasing how technology can contribute to a more just and balanced world.

These ethical practices, rooted in human welfare, environmental stewardship, transparency, and equity, position the infrastructure as a global resource governed by principles that transcend technological ambition. By upholding these values, the infrastructure sets a standard for responsible innovation, demonstrating that humanity's greatest advancements can and should serve the highest ideals. This ethical framework not only safeguards the infrastructure's legacy but also inspires a future where technology is consistently guided by a vision of shared prosperity and respect for the natural world.

As the infrastructure for space-based energy reaches its full potential, it promises to fundamentally transform the world by making universal access to energy a reality. This vision extends beyond simply meeting energy needs; it offers a path to a society where opportunities for innovation, personal development, and cultural enrichment are no longer constrained by resource scarcity. In this world, access to abundant, clean energy enables humanity to unlock its fullest potential, breaking free from the limitations that have historically defined progress and prosperity.

A future where energy scarcity no longer exists reshapes industries, allowing them to operate at new levels of efficiency and sustainability. Agriculture, manufacturing, and transportation—sectors traditionally constrained by energy costs—can evolve toward models that embrace renewable practices without compromise. Agriculture, empowered by limitless energy, could support urban vertical farms and vast irrigation systems that address global food security, transforming regions once reliant on imports into thriving agricultural hubs. In manufacturing, energy-intensive processes such as material recycling and sustainable production become economically viable, reducing environmental impact and fostering a circular economy. Transportation, now driven by abundant clean energy, facilitates efficient, low-emission logistics networks, connecting people and goods across continents without reliance on fossil fuels. Each industry benefits, contributing to a society where economic growth is no longer in conflict with environmental stewardship.

On a personal level, the impact of universal energy access permeates daily life, empowering individuals and communities with opportunities for growth and resilience. Energy access becomes a cornerstone of modern infrastructure, delivering reliable power to homes, schools, and hospitals, even in remote or previously underserved areas. The infrastructure for healthcare expands, allowing medical facilities to operate with the necessary energy resources for advanced treatments, refrigeration of essential medicines, and 24/7 lighting and climate control. Education, too, flourishes, as schools equipped with digital resources reach students worldwide, bridging educational divides that were once exacerbated by geography and resource scarcity. With energy no longer a limiting factor, individuals and communities find themselves equipped to innovate, learn, and thrive, creating a ripple effect that extends beyond immediate benefits to shape generations.

The infrastructure's environmental impact also heralds a new chapter for planetary health. As fossil fuel dependency wanes, the natural world experiences a resurgence, with reduced emissions and revitalized ecosystems. Large-scale environmental projects, powered by abundant clean energy, address long-standing ecological damage: reforestation efforts restore biodiversity, carbon capture systems mitigate atmospheric carbon levels, and water desalination plants provide fresh water to arid regions. With energy as a regenerative force, humanity gains the tools to restore balance to the planet's ecosystems, reversing centuries of environmental degradation and fostering a harmonious relationship between technological progress and ecological preservation.

Perhaps most profoundly, energy abundance redefines humanity's approach to innovation and creativity. Freed from the burdens of resource limitation, society is empowered to explore pursuits that expand intellectual, cultural, and spiritual horizons. Artists, researchers, scientists, and visionaries have the freedom to pursue groundbreaking work without the constant constraint of financial viability tied to resource costs. This freedom enriches society as a whole, encouraging diverse forms of expression and discovery that deepen our collective understanding and enhance the human experience.

Through this transformative infrastructure, humanity steps into a future defined not by scarcity but by the potential to create, collaborate, and grow in unprecedented ways. This vision challenges traditional notions of power and resource allocation, inviting a new paradigm where prosperity is measured not by material wealth but by the shared potential unlocked within every individual and community. In this era of energy abundance, humanity reclaims its ability to envision and pursue a world that honors the highest aspirations of progress, sustainability, and shared prosperity.

Conclusion: A Blueprint for Humanity's Energy Future

As humanity approaches the cusp of a revolutionary era in energy, we find ourselves armed with unprecedented opportunities to transcend the limitations that once defined our world. The vision of energy compounding stands as both a profound scientific breakthrough and an ethical beacon, presenting a new paradigm where power is neither a fleeting resource nor a privilege, but a regenerative force amplifying itself with each cycle. This principle—drawing on the wisdom of natural systems and the precision of finance—ushers in a future where the boundless power of zero-gravity environments and superconductive materials converge, creating a self-sustaining network of energy that not only meets but exceeds global demand.

In this journey from scarcity to abundance, humanity rediscovers the foundational role of energy not merely as a commodity but as the lifeblood of civilization's aspirations. For centuries, scarcity has molded our economies, structured our societies, and confined our ambitions. The reliance on fossil fuels, the promise and limitations of renewables, and the continuous tug-of-war between consumption and conservation have restricted our collective progress. But today, as we step into a world where energy systems regenerate themselves, the shadow of scarcity fades, replaced by the illuminating prospect of sustainable abundance.

The compounding model of energy production transforms energy from a linear to an exponential resource, mirroring the principles of compounding interest in finance and the cyclical regeneration of ecosystems. In zero-gravity settings, where resistance is minimized, and superconductive technologies eliminate energy loss, we establish a framework that not only meets our immediate needs but reinvests every surplus, enabling a cycle of growth that is virtually infinite. This vision calls for a departure from traditional resource exploitation and invites us to embrace a regenerative model where energy expands with each use, aligning technological innovation with humanity's most pressing needs.

By positioning energy-compounding stations at stable Lagrange points in space, we harness the stability of celestial mechanics to maintain a seamless flow of power, devoid of terrestrial limitations. Here, energy generation and storage operate in an environment optimized for efficiency, drawing power from the universe itself. This network of orbital stations, equipped to transmit power wirelessly back to Earth, promises to reimagine the very infrastructure of energy distribution, freeing it from geographical and political constraints. Energy, once bound by grids and grids of cables, becomes a universal resource, accessible across continents, empowering economies, and uplifting communities.

This era of energy abundance extends far beyond technological achievement; it reshapes society, transforming the way we approach challenges and catalyzing a renaissance of creativity, exploration, and stewardship. For the first time, communities in regions historically marginalized by energy poverty gain the foundation for sustainable development, as reliable power sparks advancements in healthcare, education, and industry. Environmental restoration efforts receive the support they need to reforest lands, rehabilitate ecosystems, and restore biodiversity. In this future, energy abundance becomes a catalyst for healing, offering the resources to combat climate change and preserve the planet for generations to come.

As we progress toward this vision, the ethical path forward becomes increasingly vital. Energy abundance, if monopolized, risks replicating the inequities of the past on a grander scale. Thus, governance frameworks grounded in transparency, accountability, and inclusivity are imperative to ensure this transformative power remains a public good, stewarded collectively for the benefit of all. Such governance honors not only the practical need for equitable access but also the philosophical commitment to fairness, as humanity seeks to harmonize technological progress with social justice and ecological responsibility.

Energy compounding redefines progress, liberating humanity from the constraints of scarcity and unlocking an era where the potential for growth, collaboration, and compassion know no bounds. This future challenges us not only to innovate but to reimagine our place in the universe, shaping a legacy of responsible advancement that honors both our aspirations and our planet. In the landscape of boundless energy, we find not just the power to sustain ourselves but the foundation upon which to build a civilization that thrives in harmony with itself and the world.

With the promise of limitless, self-sustaining energy, humanity stands poised to transcend the competitive paradigms of resource acquisition that have defined history. In a world where energy flows not as a restricted commodity but as a universal right, the emphasis shifts from survival to the cultivation of human potential. Freed from the constraints of scarcity, societies can pivot toward addressing the deeper issues that have long hindered progress: inequality, environmental degradation, and global connectivity. The liberation from finite resources marks a pivotal evolution, one where progress is not bounded by economic competition but thrives on collaboration, creativity, and the shared pursuit of knowledge.

The societal transformation brought about by energy abundance reaches into the heart of every sector. Healthcare systems, unburdened by the cost and limitations of power, can expand their reach, providing life-saving technology and preventive care to even the most remote populations. Education, once limited by access to resources, becomes an immersive, universal experience, where every child and adult has the opportunity to learn and innovate. Agricultural systems, strengthened by reliable energy, adopt practices that are both sustainable and scalable, ensuring food security and reducing reliance on finite resources. In this future, energy abundance fuels the infrastructure of human development, turning challenges into opportunities and creating a foundation for inclusive prosperity.

Yet, as power becomes universally accessible, the ethical responsibilities tied to this transformation grow equally profound. This shift necessitates a governance structure that preserves energy as a common good, immune to the monopolistic tendencies that have historically plagued resource distribution. Transparent, international cooperation must govern the deployment and oversight of energy infrastructure, establishing a framework that guarantees fair access while safeguarding against exploitation. In this framework, energy abundance serves not as a tool for power or profit but as a testament to humanity's commitment to equitable progress, shaping a global society founded on mutual respect and shared responsibility.

Such ethical considerations extend into the environmental realm, where the abundance of energy enables humanity to prioritize restoration over extraction. The legacy of pollution and habitat destruction tied to fossil fuel dependence is gradually reversed as sustainable practices replace environmentally harmful processes. With a consistent, renewable power source, large-scale projects to reclaim ecosystems, replenish forests, and support biodiversity gain momentum. This approach redefines humanity's relationship with nature, fostering a model of growth that harmonizes technological advancement with ecological balance. In this era, energy abundance empowers not only human aspirations but also the resilience of the natural world, supporting a future where human ingenuity and environmental stewardship are intertwined.

Central to this transformation is the principle of autonomy, where communities are equipped with the resources and knowledge needed to manage their own energy systems. This autonomy strengthens local economies, empowering regions to thrive independently and develop sustainably on their terms. In providing access to energy, we offer a pathway to self-determination, enabling societies to cultivate resilience from within. The result is a global landscape where progress is decentralized, inclusive, and sustainable—a mosaic of empowered communities linked by a shared foundation of energy abundance.

In this new paradigm, the concept of progress shifts from resource consumption to ethical advancement, measured not by the accumulation of wealth but by the quality of life and environmental health. The abundance of energy fosters a culture where innovation is driven by the potential for positive impact rather than profit, transforming industries, social structures, and individual aspirations. Science, art, and technology converge in a society that values shared progress, where each achievement contributes to a legacy of sustainable growth and intellectual exploration.

This vision culminates in a redefined world where energy abundance is the bedrock of a global civilization that prioritizes human welfare, environmental stewardship, and ethical responsibility. Freed from the limitations of finite resources, humanity is invited to pursue a new form of exploration—one that seeks to understand and protect the interconnectedness of life on Earth and beyond. This future is not merely a product of technological innovation but a testament to the collective resolve to build a civilization that honors the principles of equity, sustainability, and stewardship.

As we look to the horizon of this energy-abundant world, the call to action becomes clear. To realize this vision, we must embrace a commitment to responsible innovation, ensuring that the benefits of energy abundance reach every individual and community. This journey requires a global coalition of leaders, innovators, and citizens, each dedicated to the ethical stewardship of this transformative resource. Together, we can create a world where the legacy of energy abundance transcends material gain, leaving a foundation of progress that serves both humanity and the planet for generations to come.

Epilogue: The Dawn of an Era

As we stand at the threshold of a new era in energy and innovation, the world finds itself on the brink of a transformation unprecedented in scope and significance. The dawn of this era is not marked solely by advancements in technology, but by a profound shift in humanity's relationship with energy itself. Where once we were bound by the constraints of limited resources, scarcity-driven competition, and environmental degradation, we now face a future replete with boundless possibilities. At the heart of this shift lies the concept of energy abundance—an inexhaustible, self-sustaining supply of clean, renewable power capable of transcending the limitations that have long dictated the terms of human advancement.

This vision of energy abundance represents more than a technological breakthrough; it signifies a paradigm shift in how humanity defines progress, prosperity, and responsibility. The journey toward this vision has been one of relentless innovation, inspired by pioneers who dared to imagine a world untethered from fossil fuels and free from the ecological and social costs of extractive practices. These pioneers have paved the way for a new generation of innovators who now stand ready to implement the systems, principles, and ethical frameworks necessary to create a world where energy scarcity is a relic of the past.

In embracing energy abundance, humanity is poised to redefine the parameters of development. Clean, sustainable power, generated through revolutionary technologies and compounded in zero-gravity environments, will provide a foundation for growth that is not only exponential but regenerative. This energy will fuel infrastructure, empower communities, and enable societies to rise above the constraints that have historically dictated resource allocation. It promises to foster a world where equality, environmental stewardship, and shared prosperity are not merely aspirations but realities made possible through the ethical deployment of boundless energy.

The impact of these pioneering breakthroughs extends far beyond technological innovation; they resonate deeply with humanity's most enduring values and aspirations. Energy abundance offers an unprecedented opportunity to bridge divides, uplifting those who have long been marginalized by energy scarcity and enabling them to participate fully in a global economy powered by sustainable resources. As nations and communities rise from the limitations of their resource-dependent pasts, they contribute to a balanced and interconnected global landscape where energy no longer dictates access or opportunity.

For the environment, the shift to sustainable energy holds equally transformative implications. By moving away from fossil fuels and embracing a model that operates with minimal ecological impact, humanity can embark on a path of restoration rather than degradation. The availability of clean energy will empower large-scale efforts in reforestation, habitat protection, and pollution reduction. This regenerative approach to energy opens doors to restore ecosystems, preserve biodiversity, and revitalize the earth's natural wealth, fostering a harmonious balance between technological progress and ecological preservation.

Such profound changes are not without their challenges, but history provides a rich tapestry of lessons in innovation and adaptation. From the industrial advances of the 19th century to the digital revolution of the late 20th century, every technological leap has demanded a recalibration of societal norms, governance structures, and ethical frameworks. These lessons guide our present efforts, underscoring the need to approach energy abundance with a commitment to responsible stewardship, transparency, and equitable access. We must learn from the past to ensure that the benefits of this new energy era are universally shared and ethically managed.

One of the most compelling examples of innovation aligned with ethical ambition is found in the work of Nikola Tesla, who envisioned a world where energy could flow freely, transcending borders and economic divisions. Although his vision was never fully realized in his lifetime, Tesla's legacy endures as a reminder of what is possible when technology is harnessed for the common good. Today, his dream of universal energy access finds renewed relevance in the development of wireless energy transmission systems that could beam power from zero-gravity stations to earthbound receivers. This technology, driven by advances in resonant inductive coupling and superconductive materials, is not merely a tribute to Tesla's vision but a practical step toward achieving the goal he so passionately championed.

This alignment of new technology with humanity's needs requires a framework of international cooperation and shared responsibility. As energy abundance redefines resource distribution, it challenges the monopolistic tendencies that have long shaped the energy sector. To ensure that this abundance serves all of humanity, governance models must prioritize inclusivity, transparency, and accountability, creating regulatory structures that prevent exploitation and ensure equitable access to power. In doing so, we safeguard the principle that energy abundance should not be a tool for control, but a vehicle for liberation, enabling every community and nation to thrive.

Yet, the promise of energy abundance is not limited to economic or environmental benefits. It represents a reimagining of human potential, freeing individuals from the limitations imposed by scarcity and enabling them to pursue lives of creativity, education, and discovery. Freed from the constraints of survival, societies can cultivate a culture of thriving, where innovation, education, and artistic expression flourish. The energy to power these aspirations will no longer be rationed, but available to fuel the highest reaches of human potential, creating a world where achievement is defined not by the exploitation of resources, but by the enrichment of lives.

The journey to energy abundance invites humanity to envision a civilization built on principles of equity, sustainability, and shared progress. It calls upon us to become stewards of a legacy that will endure beyond our lifetimes, guiding future generations toward a world where energy serves not as a constraint, but as a catalyst for human flourishing. As we stand on the precipice of this new age, we are tasked with the profound responsibility of ensuring that each technological advance serves the greater good, safeguarding the planet and uplifting every community within it.

In a world where energy scarcity becomes obsolete, humanity's approach to development can transcend the historical conflicts and competitions driven by finite resources. This new era is shaped by the principles of abundance, where nations no longer need to compete for energy dominance or to control the critical resources that once dictated power dynamics. Instead, the energy-compounding model creates a foundation for shared prosperity, where advancements in one region can ripple outward, empowering and uplifting societies on a global scale. This vision of shared energy abundance has the potential to redefine the relationships among nations, fostering a spirit of interdependence rather than rivalry, and guiding the world toward a collective goal of universal prosperity.

At the core of this shift lies the principle of regenerative growth, a process that mirrors the cycles of nature where energy flows continuously, supporting growth without depletion. Just as ecosystems thrive by recycling nutrients in an unending loop, energy compounding envisions a world where each unit of generated power sustains and amplifies itself. This regenerative model transcends the limitations of traditional consumption, replacing extraction with renewal and scarcity with abundance. By adopting this self-sustaining cycle, humanity moves from a linear model of use and waste toward a circular system where energy not only meets immediate needs but expands in harmony with the collective ambitions of society.

Realizing this vision requires more than technological breakthroughs; it demands a philosophical transformation in how humanity conceives of energy and its role in civilization. Energy abundance offers a foundation upon which a new economic model can emerge—one that breaks free from the cycles of extraction, depletion, and waste that have characterized industrial economies. In this model, growth is not defined by the consumption of finite resources but by the continuous expansion of renewable, regenerative energy that supports innovation and ecological restoration. With an abundance of clean, sustainable power, societies are free to invest in long-term solutions that foster resilience, environmental health, and equitable prosperity.

This new economic model also challenges the notion of scarcity as an immutable condition of human existence. Throughout history, scarcity has been the silent force shaping economies, policies, and even cultural identities. From ancient agrarian societies to modern industrial nations, the availability of resources has defined the bounds of possibility, restricting both the ambitions of individuals and the growth of entire societies. But with the advent of energy abundance, humanity has an opportunity to redefine progress in terms that align with both ethical responsibility and ecological balance. In this world, energy abundance enables the realization of potential that was previously unimaginable, shifting the focus from mere survival to the pursuit of a thriving, sustainable future for all.

The environmental implications of energy abundance cannot be overstated. As society shifts away from fossil fuels and embraces renewable, compounded energy sources, humanity gains the capacity to restore ecosystems rather than exploit them. Large-scale environmental restoration projects, supported by inexhaustible clean energy, become not only feasible but also scalable, allowing societies to rehabilitate degraded lands, restore marine habitats, and reestablish biodiversity on a global scale. This shift toward regenerative environmental practices signifies a new chapter in humanity's relationship with the natural world, one where technological progress and ecological health are not in opposition but are harmonized to create a sustainable legacy.

In this regenerative model, energy abundance supports initiatives that prioritize ecological balance and preservation. For example, vast reforestation projects could counteract decades of deforestation, creating carbon sinks that absorb greenhouse gases and help stabilize the climate. Similarly, clean energy can drive the desalination of seawater, bringing fresh water to arid regions and transforming agriculture in areas historically constrained by water scarcity. These projects, once seen as ambitious or even utopian, become practical and achievable within the framework of energy abundance, illustrating how a regenerative approach to energy can serve as a catalyst for both environmental restoration and social progress.

However, the promise of energy abundance also brings new responsibilities, especially in managing this resource equitably and sustainably. As history has shown, advancements in technology and energy have the potential to uplift societies but can also create disparities if not governed with foresight and fairness. The Industrial Revolution, though a period of tremendous economic growth, also produced environmental degradation and widened economic divides that persist to this day. To avoid repeating these patterns, the adoption of energy-compounding technology must be accompanied by ethical frameworks that prioritize inclusivity, accountability, and environmental stewardship. Such frameworks ensure that energy abundance does not concentrate power in the hands of a few but remains a resource accessible to all, fostering a world where innovation and equity coexist.

The governance of energy abundance will necessitate a new form of collaboration, one that transcends national borders and prioritizes the collective well-being of humanity. International agreements, guided by principles of fairness and sustainability, will be essential in establishing a regulatory framework that prevents monopolization and ensures that energy abundance remains a shared resource. Public institutions, private sectors, and civic organizations must work together to create policies that protect this resource from exploitation and promote its equitable distribution. Such governance structures represent not only a commitment to responsible resource management but also a dedication to upholding the values of transparency, inclusivity, and shared prosperity.

In education, the promise of energy abundance extends far beyond the classroom, empowering students to learn, experiment, and innovate in ways previously limited by resource constraints. Schools in remote or underserved areas, once hindered by unreliable power and limited infrastructure, would have access to consistent, high-quality energy, enabling them to implement advanced digital tools and interactive learning experiences. With abundant energy, educators can introduce new technologies, expand curriculums, and foster environments that encourage creativity, critical thinking, and scientific inquiry. This educational transformation would prepare the next generation not only to navigate a world of energy abundance but to continue advancing its principles, ensuring that each generation builds upon the progress of the last.

Furthermore, energy abundance invites a reimagining of public health and community well-being. Hospitals, clinics, and healthcare facilities, especially those in underserved areas, would gain reliable access to the power necessary to operate life-saving equipment, store medical supplies, and expand services. Communities would benefit from enhanced public health infrastructure, supported by an energy supply that does not falter in times of need. This reliable energy access would enable healthcare systems to respond effectively to crises, reduce mortality rates, and improve quality of life across all demographics. In this model, energy abundance becomes a vital component of public health, enabling healthcare providers to serve communities equitably and efficiently.

Ultimately, the vision of energy abundance extends far beyond the immediate benefits of clean and sustainable power; it represents an invitation for humanity to rethink the very foundations of society. The shift from scarcity to abundance challenges long-held assumptions about progress, prosperity, and purpose, encouraging societies to explore new possibilities for cultural enrichment, scientific advancement, and global unity. In this world, energy is no longer a commodity to be fought over, but a resource that empowers, connects, and uplifts every corner of the earth.

As humanity embarks on this journey, it is crucial to remember that the power of energy abundance lies not only in its technological capabilities but in its capacity to foster a world where progress aligns with principles of equity, stewardship, and sustainability. This future invites all individuals and communities to participate in shaping a civilization where energy abundance serves as both a foundation and a guiding light, illuminating the path toward a world where human potential is no longer constrained by the limitations of the past but is liberated to reach unprecedented heights.

As humanity steps forward into a world fueled by limitless energy, we confront not only the vast possibilities for growth and innovation but also the profound responsibility to wield this power with wisdom. This responsibility is reflected in the ethical and philosophical foundations that must accompany the technological revolution of energy abundance. To cultivate a society that fully benefits from this new era, we must ensure that the principles guiding this transformation are rooted in justice, equity, and sustainability, transcending the narrow interests of profit or control. By building these principles into the very framework of energy governance, humanity can shape a legacy that honors the values of shared prosperity, environmental stewardship, and respect for human dignity.

Central to this ethical vision is the commitment to universal access to energy—a principle that aligns with the belief that access to power is not a privilege but a right. For centuries, energy has been a defining factor in the development and prosperity of nations, creating stark divides between those who have abundant resources and those who struggle with scarcity. In this new era, the principles of equity and justice call for a system that eliminates such divides, ensuring that every individual, regardless of location or socioeconomic status, has access to clean, reliable, and affordable energy. This transformation of access would dismantle historical inequalities and empower communities across the globe to build resilient infrastructures, cultivate economic growth, and enhance quality of life.

As energy abundance becomes the norm, the democratization of power distribution will necessitate an international framework of governance, grounded in the principles of transparency and accountability. Such a framework would protect against monopolistic control, ensuring that no single entity, whether corporate or governmental, can dominate this critical resource. To achieve this, international coalitions and regulatory bodies must work together to establish policies that promote fair distribution, prevent exploitation, and empower local communities to have a say in their energy futures. This shared governance model embodies a commitment to the idea that energy abundance is a collective asset, one that belongs not to any one country or corporation but to humanity as a whole.

In addition to universal access, the ethical deployment of energy abundance requires that societies prioritize the sustainability of their practices. Although energy may no longer be a limited commodity, the infrastructures and ecosystems it touches must be protected to ensure long-term environmental health. To that end, the principles guiding energy abundance must embrace a regenerative approach to resource management—one that not only meets the needs of the present but actively works to restore and enhance the natural world for future generations. This regenerative philosophy represents a departure from the extractive models of the past, advocating instead for a system where technological advancement harmonizes with ecological preservation.

The alignment of technological progress with environmental ethics also creates a fertile ground for innovation in the field of environmental restoration. With energy abundance, humanity can undertake projects once considered impossible or prohibitively expensive. The power to desalinate seawater on a massive scale, for example, could provide freshwater to drought-stricken regions, transforming arid landscapes into fertile grounds for agriculture and habitation. Likewise, the capability to reforest degraded lands on an unprecedented scale would allow humanity to address climate change more directly, absorbing carbon dioxide and restoring biodiversity. These projects are not merely side benefits of energy abundance; they are central to the vision of a civilization that respects and nurtures the natural world upon which all life depends.

While the potential applications of energy abundance are boundless, the true impact of this new era depends on our commitment to instill an ethos of responsibility in future generations. Education will play a vital role in fostering a global culture that values sustainability, stewardship, and ethical innovation. By integrating these principles into curricula at all levels, societies can ensure that the next generation of leaders, scientists, and innovators are equipped not only with technical skills but with a profound sense of duty to the collective well-being of humanity and the planet. In classrooms around the world, students would learn not only the science of energy but also the ethics of its use, preparing them to steward this resource with wisdom, integrity, and a commitment to equity.

In a world of energy abundance, educational institutions become hubs of innovation and ethical learning, nurturing a generation that sees beyond short-term gains to envision a future where every advancement contributes to the common good. Schools in remote or underserved areas, once constrained by limited resources, will have access to the power necessary to create state-of-the-art learning environments. This democratization of knowledge, supported by the pillars of energy abundance, will enable young minds to explore the realms of science, technology, and creativity without the boundaries that once limited their predecessors. Such an environment fosters not only individual achievement but collective growth, creating a global community of learners, thinkers, and creators empowered to advance the ideals of a sustainable and equitable world.

The social and cultural impacts of energy abundance extend beyond education, reshaping the very fabric of human society. Communities that once struggled with unreliable power and limited resources will experience a renaissance in cultural and social expression. Artists, writers, and thinkers will find new platforms and support for their work, creating a flourishing of creativity that enriches human experience. Digital technology, enabled by an inexhaustible power supply, will connect individuals across vast distances, facilitating cross-cultural exchanges that bridge divides and celebrate the diversity of human expression. In this way, energy abundance does more than fuel physical progress; it fosters a global renaissance in culture, creativity, and collaboration.

In healthcare, the availability of abundant, clean energy will revolutionize the ability of medical institutions to deliver high-quality care. Hospitals and clinics, particularly those in underserved areas, will have reliable power to operate advanced medical equipment, maintain critical infrastructure, and expand services to reach more patients. Medical research facilities will also benefit, enabling advancements in disease treatment, genetic research, and preventive care, supported by continuous power to fuel life-saving innovations. This transformation of healthcare, powered by abundant energy, has the potential to extend life expectancy, reduce mortality, and improve quality of life across all demographics, underscoring the principle that energy abundance should serve as a foundation for human well-being.

As energy abundance reshapes society, it also transforms the role of government and public institutions, calling for new forms of leadership that prioritize transparency, inclusivity, and ethical accountability. Governments will play a crucial role in managing this resource, implementing policies that promote fair distribution and prevent exploitation. Public institutions, empowered by an unwavering commitment to the common good, must safeguard energy abundance from becoming a tool for control or a means of exclusion. By championing inclusive policies and fostering partnerships with private sectors, governments can ensure that energy abundance fulfills its promise to uplift all communities, creating a society where power, in every sense, is shared.

To achieve this vision, the global community must embrace a new philosophy of progress—one that views success not merely in terms of technological achievement but in the creation of a society that values compassion, stewardship, and responsibility. In this reimagined world, energy abundance is a means to an end, a resource that allows humanity to transcend the limitations of the past and build a civilization that prioritizes human dignity, environmental health, and social equity. This philosophy of progress encourages humanity to redefine its ambitions, focusing not on the accumulation of resources but on the enrichment of lives, the preservation of ecosystems, and the empowerment of every individual.

The realization of energy abundance represents more than a technological evolution; it is a moral awakening, inviting humanity to reconsider what it values and how it can achieve these values in a way that honors both people and planet. This moment in history calls upon us to choose a path that aligns with our highest principles, to view energy not as an end in itself but as a catalyst for creating a society that honors the dignity of all its members. By embracing this ethos, we lay the foundation for a legacy that will endure, a world where every individual has the opportunity to thrive in a society that respects both human potential and the natural world.

As we look toward the future, the journey of energy abundance invites every generation to contribute to its realization. This journey will not be without its challenges, but with each step forward, we bring humanity closer to a world where energy empowers and uplifts, where progress and preservation go hand in hand, and where the legacy of abundance serves as a beacon of hope for generations to come. The path ahead is illuminated by a profound vision—a vision of a civilization that transcends scarcity, embraces sustainability, and cultivates a future worthy of the boundless potential that energy abundance makes possible.

The profound promise of energy abundance brings with it an opportunity to reimagine civilization itself—a society that transcends survival to focus on thriving. Freed from the constraints of limited resources, humanity can shift its collective energy toward pursuits that enhance the human experience, cultivate resilience, and foster a deeper connection with the natural world. In this vision, energy abundance is not merely a technological achievement but a transformative force that reshapes the foundations of culture, governance, and individual potential, illuminating a path where humanity no longer measures progress by what it consumes but by what it creates and preserves.

One of the most significant impacts of this transformation lies in the redefinition of human potential. With energy as an abundant and accessible resource, individuals are empowered to explore their fullest capacities, unbound by the limits of scarcity. Creativity, innovation, and intellectual pursuits flourish in a world where energy enables rather than constrains. Freed from the burdens of energy insecurity, people across all communities can dedicate their time and resources to endeavors that deepen understanding, foster compassion, and propel humanity forward in ways previously unimaginable. This cultural shift, supported by the power of energy abundance, invites a collective renaissance in which society places equal value on both scientific advancement and artistic expression, on technological prowess and ethical progress.

In this world, energy abundance redefines the parameters of personal and societal development. Education becomes a universal right supported by a steady supply of power, ensuring that even the most remote or impoverished communities have access to quality resources. Schools evolve into centers of discovery where students engage in hands-on experimentation, critical thinking, and collaborative problem-solving. Knowledge, once limited by geographic or economic boundaries, flows freely, enabling all people to learn, grow, and contribute to a shared pool of understanding. This educational empowerment creates a ripple effect, producing generations of informed, capable individuals equipped to address the complex challenges of their time and to carry forward the legacy of sustainable abundance.

At the heart of this cultural and educational evolution is a renewed emphasis on ethical responsibility. The availability of boundless energy invites not only innovation but also introspection, urging humanity to consider how best to steward this resource for the collective good. In a world no longer constrained by scarcity, the focus of progress shifts from competition over resources to a shared commitment to ethical use and preservation. This ethos of responsibility is embedded in every level of education, from primary schools to universities, where students are encouraged not only to master the scientific principles of energy but to understand the ethical frameworks that govern its use. In this way, the next generation is not only technologically proficient but morally grounded, prepared to uphold the values of stewardship and equity that define a world of abundance.

The effects of energy abundance extend into every facet of life, fundamentally altering how individuals, communities, and nations relate to one another. Freed from the pressures of resource competition, societies can cultivate relationships built on collaboration and mutual respect. Nations that once vied for dominance over oil, coal, or minerals find common ground in a shared commitment to sustainable growth, fostering alliances that transcend historical divisions. International cooperation, motivated by the collective benefits of abundant energy, lays the foundation for a new global order, one that prioritizes environmental preservation, human rights, and shared prosperity over territorial or economic conquest. In this world, diplomacy is guided not by scarcity-driven politics but by a commitment to fostering a harmonious and resilient global community.

Energy abundance also transforms how humanity interacts with the natural world. No longer dependent on extractive practices that degrade ecosystems, societies can adopt regenerative approaches that restore, protect, and celebrate biodiversity. With clean, sustainable energy powering every sector, industries can operate without the environmental toll associated with fossil fuels, opening avenues for conservation on an unprecedented scale. The mining of precious minerals, deforestation for fuel, and the pollution of water and air give way to practices that honor the planet's ecosystems, recognizing their intrinsic value and essential role in supporting all life.

In this regenerative model, humanity becomes a true steward of the earth, harnessing energy not to exploit but to sustain and rejuvenate. Projects once constrained by the limits of conventional energy sources—such as large-scale reforestation, ocean clean-up initiatives, and soil restoration—become practical and scalable, empowered by a consistent supply of clean power. By investing in the health of natural systems, humanity ensures that the earth can continue to support diverse life forms, creating a legacy of environmental resilience that will endure for generations. This harmony between technological advancement and ecological preservation embodies a new vision for civilization, one where human progress is measured not by what it takes from the earth but by what it gives back.

Beyond environmental stewardship, the ethical imperatives of energy abundance call for an approach to governance that prioritizes transparency, equity, and inclusion. As energy becomes a universally accessible resource, the role of governments and public institutions evolves to support and protect this shared asset. Policies that promote fair distribution, prevent monopolization, and ensure that energy abundance remains accessible to all become pillars of governance, reinforcing the principle that energy is a public good, not a private commodity. This approach represents a shift from the hierarchical, scarcity-driven models of the past toward a more egalitarian structure, where decision-making processes are inclusive and communities have a voice in how energy resources are managed and distributed.

In this vision of governance, public and private sectors work together to create regulatory frameworks that encourage innovation while safeguarding the public interest. Transparency in decision-making, coupled with accountability to the communities served, ensures that energy abundance fulfills its promise of equity and inclusivity. International organizations, governments, and civic institutions form coalitions dedicated to maintaining the ethical integrity of energy resources, recognizing that the benefits of this abundance are most meaningful when they are shared. This commitment to inclusive governance fosters trust, empowering societies to embrace energy abundance as a foundation for sustainable growth, cultural flourishing, and global unity.

The power of energy abundance to unify humanity also extends to the realms of art, culture, and human connection. As societies move beyond the struggle for survival, they gain the freedom to celebrate the richness of human creativity, to cultivate spaces where art, music, literature, and philosophy can thrive. Abundant energy supports the digital platforms that connect people across continents, creating a virtual tapestry of ideas, stories, and expressions that celebrate both the diversity and unity of human experience. Cultural exchange flourishes, fostering a world where differences are not sources of division but inspirations for collaboration and mutual respect.

This cultural renaissance, supported by the democratization of energy, allows individuals from all backgrounds to contribute to a global conversation, enriching human understanding and empathy. Art and expression become accessible to all, no longer constrained by limited resources or geographical isolation. In this world, creativity is celebrated as a universal human right, a source of joy, reflection, and connection that transcends boundaries. By nurturing these connections, energy abundance helps to build a world where cultural understanding is deepened, communities are strengthened, and the collective consciousness is elevated, creating a society that values both individuality and interconnectedness.

The promise of energy abundance also redefines the role of the individual within society. In a world where energy is freely available, individuals are empowered to pursue their passions, innovate fearlessly, and contribute meaningfully to their communities. Entrepreneurship flourishes as people are no longer held back by the constraints of limited power or the high costs of infrastructure. Small businesses, local artisans, and social entrepreneurs gain the resources they need to bring their ideas to life, fostering economic diversity and resilience. This empowerment at the individual level creates a dynamic society where everyone has the opportunity to shape their future, contributing to an economy that is as innovative as it is inclusive.

Ultimately, the journey toward energy abundance is not merely a technological revolution but a philosophical evolution. It invites humanity to reexamine its values, to consider what it means to progress in harmony with nature and each other. This vision of abundance is rooted in the belief that human potential is limitless, that societies flourish not through competition but through collaboration, and that true progress honors the planet that sustains us all. In embracing this vision, humanity steps into a new era defined not by scarcity but by possibility—a future where energy abundance fuels the aspirations of every individual, supports the health of the natural world, and unites all people in a shared commitment to a thriving, sustainable, and just civilization.

As we envision this future, the legacy of energy abundance calls upon us to be vigilant stewards, to cultivate a world where power empowers rather than divides, where every technological advance serves the greater good, and where the collective potential of humanity is harnessed to create a civilization worthy of the boundless energy at its disposal. This is the dawn of a new era, one that transcends the achievements of the past and sets a course for a world where energy is not only a resource but a catalyst for unity, compassion, and hope—a foundation upon which the dreams of future generations will be built.

In this vision of a world transformed by energy abundance, society embarks on a path where progress aligns with a deepened understanding of our shared responsibilities. This transformation extends beyond technology, beyond the structures of governance, and into the very heart of what it means to be human. As individuals and communities embrace the potential of energy abundance, they are invited to redefine prosperity, to think not in terms of accumulation but of contribution, to view success not as a zero-sum game but as a collective ascent toward a world enriched by shared goals and mutual respect. This era brings into focus a new paradigm of human development—one that celebrates collaboration, fosters resilience, and upholds the dignity of every individual.

In this reimagined society, the principles that guide energy abundance also inform social frameworks, shaping a culture that values interdependence as a source of strength. Freed from the specter of scarcity, communities are encouraged to invest in one another, to share knowledge and resources openly, and to build networks of support that transcend traditional boundaries. This spirit of unity creates an environment where local, national, and global initiatives can flourish side by side, each reinforcing the other in a shared commitment to the common good. The result is a world that is not merely connected but interwoven, where the triumphs of one community resonate across continents, inspiring others to pursue their own visions of progress and well-being.

Education, as a pillar of this society, evolves into a lifelong journey available to all. With energy no longer a limiting factor, educational institutions expand their reach, creating inclusive, adaptable learning environments that cater to the needs of every individual, regardless of age, location, or background. Schools become hubs of curiosity and discovery, where students are not simply taught but are encouraged to question, to explore, and to innovate. The curriculum expands to reflect the ethical and ecological imperatives of the new era, teaching not only the sciences of energy and technology but also the values of empathy, stewardship, and social responsibility. This holistic approach prepares individuals not only to thrive within the systems of energy abundance but to contribute actively to their refinement and evolution, ensuring that each generation builds upon the progress of the last.

With education as a foundation, the benefits of energy abundance permeate every aspect of society, reshaping industries and redefining economies in ways that prioritize sustainability and inclusivity. Traditional industries, once constrained by resource limitations and environmental costs, find new pathways to innovation as they adapt to the possibilities of abundant energy. Manufacturing, agriculture, and transportation evolve into sectors that no longer exhaust resources but regenerate them, embracing practices that are as restorative as they are productive. In this model, economic growth is decoupled from environmental degradation, allowing industries to thrive without compromising the health of the planet. Energy abundance becomes the engine of an economy that values renewal over extraction, creating a system where progress is measured by the well-being of people and the environment alike.

In agriculture, energy abundance opens doors to practices that enhance food security and reduce environmental impact. Vertical farms, powered by clean energy, flourish in urban centers, providing fresh produce to local populations without the need for extensive transportation or soil depletion. These farms, supported by a consistent energy supply, operate year-round, offering resilient food sources that adapt to climate variability and population demands. In rural areas, clean energy enables the implementation of sustainable irrigation systems, allowing farmers to optimize water use, increase yields, and preserve biodiversity. This transformation of agriculture supports not only the nutritional needs of populations but also the ecological health of the landscapes that sustain them, fostering a harmonious relationship between food production and environmental conservation.

Transportation, too, undergoes a profound shift as energy abundance enables the widespread adoption of clean, efficient technologies. Electric vehicles, powered by renewable energy, become the standard for personal and public transit, reducing emissions and alleviating the environmental burdens associated with fossil fuels. High-speed rail networks, connected to clean power sources, link cities and regions with unprecedented efficiency, reducing the need for air travel and minimizing the carbon footprint of long-distance transport. This reimagining of mobility makes travel more accessible and sustainable, creating a transportation infrastructure that not only supports economic activity but enhances quality of life by reducing pollution, traffic, and environmental degradation.

As industries and economies transform, so too does the role of public policy, which evolves to support the ethical and equitable management of energy resources. Governments, empowered by a commitment to transparency and accountability, prioritize policies that protect public access to energy, prevent monopolization, and promote sustainable practices. The management of energy abundance requires an active, engaged approach to governance, one that anticipates potential challenges and addresses them with foresight and compassion. Public institutions, guided by the principles of stewardship, work in partnership with private sectors and international organizations to create regulatory frameworks that ensure energy remains a shared resource, accessible to all and protected from exploitation.

The shift in governance also calls for a renewed focus on human rights, recognizing that access to energy is a fundamental right that underpins many aspects of modern life, from healthcare and education to economic opportunity and environmental justice. In this world, energy abundance is viewed not as a privilege but as a necessity, one that enables individuals to live with dignity, freedom, and opportunity. Public policies reflect this understanding, striving to create a society where every person has the resources they need to pursue their goals, contribute to their communities, and lead fulfilling lives. This approach to governance strengthens the social fabric, creating a foundation of trust and unity that supports the sustainable management of energy abundance for generations to come.

In this transformed world, the focus on equity and inclusion extends to international relations, fostering a global community where nations work together to address shared challenges and pursue common goals. Energy abundance enables a form of diplomacy that prioritizes cooperation over competition, creating opportunities for countries to collaborate on issues such as climate change, poverty alleviation, and technological innovation.

By sharing knowledge, resources, and expertise, nations contribute to a global network of support, where advancements in one region benefit the entire world. This spirit of cooperation lays the groundwork for a new era of international unity, one that transcends traditional power dynamics and embraces a shared commitment to planetary well-being.

The ethical imperatives of energy abundance also create a foundation for social resilience, equipping communities to face future challenges with confidence and adaptability. With reliable access to clean energy, communities can build infrastructures that withstand natural disasters, climate impacts, and economic fluctuations. Emergency response systems, powered by abundant energy, are capable of providing immediate support to affected areas, ensuring that resources are available when and where they are needed most. This resilience extends beyond physical infrastructure to the social structures that bind communities together, fostering networks of mutual aid, trust, and cooperation. In this way, energy abundance not only supports material prosperity but strengthens the human bonds that sustain societies through adversity.

The vision of energy abundance ultimately redefines the aspirations of humanity, inviting us to consider what a truly prosperous society looks like in a world free from scarcity. It asks us to rethink the very nature of success, to question whether progress can be measured by wealth and consumption alone or whether it must also be gauged by the values of compassion, respect, and responsibility. In this reimagined world, prosperity is not about accumulation but about contribution; success is defined not by individual achievement but by collective well-being. This philosophy of abundance creates a society that values both individual growth and social cohesion, a world where every advancement is guided by the principles of empathy, sustainability, and shared purpose.

As we continue on this journey, the path toward energy abundance invites us to engage with our highest ideals and to consider the legacy we wish to leave for future generations. This legacy is not one of extraction or exploitation, but of stewardship and regeneration. It is a legacy that honors the earth and all its inhabitants, that respects the interconnectedness of life, and that prioritizes the well-being of both people and planet. In embracing this vision, humanity has the opportunity to build a civilization that not only meets the challenges of the present but creates a foundation of hope and possibility for the future.

The realization of energy abundance is a testament to human ingenuity, courage, and vision. It is a triumph not only of technology but of the human spirit, a reflection of our collective ability to dream, to innovate, and to strive for a better world. As we step into this new era, we are reminded that the journey of energy abundance is not a destination but an ongoing process, one that will continue to evolve as we learn, adapt, and grow. Each generation will have its role to play, contributing to the unfolding story of a civilization that transcends scarcity and embraces a future defined by unity, compassion, and resilience.

As humanity advances into this era of energy abundance, we are poised to embark on a journey not merely defined by what we can achieve but by how we achieve it. This vision calls upon us to recognize that the promise of abundance comes with a responsibility to honor the resources we now wield and to employ them for the betterment of all. In this final reflection, we confront the essence of what it means to wield limitless energy and, with it, limitless potential: to create, to sustain, and to uplift. This power is not a gift but an earned opportunity, forged through centuries of innovation and perseverance. How we use it will define not just the future of energy but the future of human civilization itself.

At the heart of this vision is a commitment to ethical stewardship. Energy abundance offers the unprecedented opportunity to reshape economies, transform industries, and lift billions out of poverty, yet it also invites us to rethink the ethical frameworks that govern these transformations. By adopting an ethos of stewardship, we ground our advancements in a philosophy that values life, dignity, and respect for both humanity and the environment. This ethos is not merely aspirational; it is foundational to creating a society that views energy not as a mere commodity but as a vital force to be managed with integrity, compassion, and foresight.

This commitment to stewardship extends beyond humanity to the entire biosphere, acknowledging that our progress is inextricably linked to the health and resilience of the natural world. With energy abundance, we have the capability to support large-scale ecological restoration efforts, to reverse environmental degradation, and to foster a symbiotic relationship with the ecosystems that sustain us. In this world, humanity becomes not an agent of consumption but a guardian of the earth, actively engaged in the preservation and regeneration of the planet's resources. This shift in perspective enables us to view ourselves not as conquerors of nature but as custodians, charged with the care and continuity of the world we inhabit.

The transformative power of energy abundance also calls for a reimagining of human aspiration. In a world where scarcity no longer defines the limits of possibility, we are free to pursue goals that reflect our highest values and our deepest hopes. With the material needs of society met, humanity can turn its focus to the enrichment of life itself—to the pursuit of knowledge, the cultivation of culture, and the exploration of the unknown. This era of abundance offers the freedom to reach beyond survival, to chart paths of innovation that expand the boundaries of human potential and that honor the diversity and richness of human experience. In doing so, we lay the groundwork for a civilization that thrives not on what it accumulates but on what it contributes to the collective journey of human advancement.

The path forward is guided by a vision of equity and unity, a world where energy abundance becomes a shared resource that empowers every community, every culture, and every individual. This vision demands that we view progress as a shared endeavor, where the benefits of technology and the fruits of innovation are accessible to all. In this world, wealth is redefined—not as the possession of finite resources but as the ability to uplift and empower others. Through this redefinition, we build a society where the success of one enriches the lives of many, and where abundance flows through networks of mutual aid and shared purpose, creating a world that is as interconnected as it is inclusive.

As we advance, we must also remain vigilant to the challenges that accompany this transformation. The transition to energy abundance will not be without obstacles, and the journey will demand resilience, adaptability, and a steadfast commitment to the values that guide us. We must guard against complacency, recognizing that the promise of abundance requires continual renewal and protection. In this world, the maintenance of energy abundance becomes an act of vigilance, a commitment to uphold the principles of equity, sustainability, and stewardship that define this era. By embracing this responsibility, we ensure that energy abundance remains a force for good, a resource that serves the common interest and inspires future generations to pursue paths of ethical innovation and responsible growth.

This vision of a civilization sustained by energy abundance transcends the limitations of individual lifetimes, inviting us to consider the legacy we wish to leave for those who will inherit the world we shape today. This legacy is not defined by monuments or achievements but by the enduring values we instill in the systems and structures we build. In embracing energy abundance, we commit to creating a world where each generation inherits not just the benefits of progress but the wisdom and responsibility to steward that progress with integrity. This generational commitment forms the foundation of a society that endures, a civilization that honors the contributions of its past while continually striving to create a better future.

As we pass this legacy forward, we entrust future generations with the tools and the principles to continue building a world that reflects humanity's highest aspirations. We empower them not only with the resources of energy abundance but with the ethical frameworks necessary to manage those resources responsibly. Through education, governance, and cultural values, we instill a commitment to equity, compassion, and sustainability, ensuring that the journey of energy abundance is one of continuous growth, adaptation, and renewal. In this way, each generation becomes both a beneficiary and a steward, a participant in a collective endeavor that spans the arc of human history and reaches toward a future of limitless possibility.

The journey toward energy abundance culminates in a vision of unity—a world where technological progress aligns with ethical purpose, where human potential flourishes alongside ecological health, and where each advancement serves the common good. This unity is not merely an ideal but a guiding principle that defines the civilization we seek to build. By embracing energy abundance as a shared resource, we create a foundation of trust and cooperation, a world where the collective strength of humanity is directed toward goals that honor life, sustain the planet, and uplift every individual. This vision of unity calls us to transcend the divides of culture, geography, and circumstance, recognizing that we are bound together by a shared commitment to a thriving, resilient, and just world.

In this dawn of a new era, humanity stands at the cusp of an unprecedented opportunity to redefine the course of its future. Energy abundance is more than a technological achievement; it is an invitation to imagine, to create, and to pursue a world where the full potential of human civilization is realized. It is a call to cultivate a society that values life, respects the environment, and celebrates the contributions of every individual. As we embrace this vision, we step into a future that transcends the limitations of the past, a future where energy empowers rather than restricts, where progress is measured not by consumption but by contribution, and where the legacy of abundance serves as a beacon of hope for all who follow.

This is the dawn of an era defined by boundless energy, boundless possibility, and boundless responsibility. As we look toward this future, let us carry forward the wisdom of those who came before, honoring the journey that brought us here and the vision that propels us onward. In this world of energy abundance, we are not just inheritors of progress; we are creators of a new story, one where humanity's greatest achievements lie not in what it takes from the earth, but in what it gives back. This story, woven from threads of innovation, compassion, and unity, stands as a testament to the power of human potential and the enduring promise of a civilization that chooses abundance over scarcity, hope over fear, and unity over division.

As we embark on this journey, we recognize that the future of energy abundance is not a conclusion but a beginning—a call to continue building, creating, and dreaming. It is a path illuminated by the vision of a world that celebrates life, that sustains the planet, and that empowers every individual to reach their fullest potential. In embracing this vision, we step forward with purpose, with courage, and with the unwavering belief that together, we can create a world that truly reflects the boundless possibilities of human civilization. This is our legacy, our responsibility, and our greatest hope—a future shaped by energy abundance and guided by the enduring values that unite us all.

www.ingramcontent.com/pod-product-compliance
Lightning Source LLC
Chambersburg PA
CBHW050645270326
41927CB00012B/2883